Compound Semiconductors
for Energy Applications and
Environmental Sustainability

MATERIALS RESEARCH SOCIETY
SYMPOSIUM PROCEEDINGS VOLUME 1167

Compound Semiconductors for Energy Applications and Environmental Sustainability

Symposium held April 14–16, 2009, San Francisco, California, U.S.A.

EDITORS:

F. Shahedipour-Sandvik
State University of New York-Albany
Albany, New York, U.S.A.

E. Fred Schubert
Rensselaer Polytechnic Institute
Troy, New York, U.S.A.

L. Douglas Bell
Jet Propulsion Laboratory
Pasadena, California, U.S.A.

Vinayak Tilak
General Electric Global Research Center
Niskayuna, New York, U.S.A.

Andreas W. Bett
Fraunhofer Institut for Solar Energy Systems
Freiburg, Germany

Materials Research Society
Warrendale, Pennsylvania

CAMBRIDGE UNIVERSITY PRESS
Cambridge, New York, Melbourne, Madrid, Cape Town,
Singapore, São Paulo, Delhi, Mexico City

Cambridge University Press
32 Avenue of the Americas, New York NY 10013-2473, USA

Published in the United States of America by Cambridge University Press, New York

www.cambridge.org
Information on this title: www.cambridge.org/9781107408258

Materials Research Society
506 Keystone Drive, Warrendale, PA 15086
http://www.mrs.org

© Materials Research Society 2009

First published 2009
First paperback edition 2012

Single article reprints from this publication are available through
University Microfilms Inc., 300 North Zeeb Road, Ann Arbor, MI 48106

CODEN: MRSPDH

ISBN 978-1-107-40825-8 Paperback

CONTENTS

COMPOUND SEMICONDUCTORS FOR ENERGY

COMPOUND SEMICONDUCTORS FOR SENSING

MATERIALS GROWTH AND CHARACTERIZATION

APPENDIX

This Appendix contains papers from
Materials Research Society Symposium Proceedings Volume **1171E**
Materials in Photocatalysis and Photoelectrochemistry for
Environmental Applications and H_2 Generation
A. Braun, P.A. Alivisatos, E. Figgemeier, J.A. Turner, J. Ye, E.A. Chandler, Editors

*Invited Paper

PREFACE

This volume contains a subset of oral and poster presentations that were made at Symposium O, "Compound Semiconductors for Energy Applications and Environmental Sustainability," held April 14–16 at the 2009 MRS Spring Meeting in San Francisco, California.

Compound semiconductors have long been an integral part of everyday life. They are a class of semiconductors that have unique properties such as a direct band gap, the ability to control the band gap and wider band gaps than seen in silicon. These properties can be leveraged for many energy-related applications such as efficient lighting, high efficiency solar cells and efficient switching. Recent progress on their potential as emitters, sensing devices in biological and chemical environments and high efficiency power devices demonstrates their impact on the conservation of energy and the environment, resulting in reduction of global warming. Compound semiconductor based photovoltaic systems are emerging as economical means of generating renewable energy through the use of concentrator technologies. Examples include energy saving solid state lighting for general illumination, and use of compound semiconductors in power switching applications such as hybrid vehicles.

Although solid state lighting devices have shown energy saving and environmental benefits, much work needs to be done to realize their full potential, including resolving the issue efficiency droop at high current densities and increasing the efficiency of LEDs with emission in green (i.e., closing the green gap). In other cases, a substantial amount of work in developing and optimizing materials and device properties may be required. Understanding the interaction of these compounds with their environment for their compatibility, and studying the potentially negative impact they may have on organisms and the natural environment are other topics that need much research. This volume contains reports from internationally known experts on the state of compound semiconductor based devices with application in environmental conservation and energy saving challenges associated with realization of such devices, and obstacles to their widespread use.

The Symposium Organizers wish to thank all who contributed to the success of this symposium, in particular the authors, reviewers, and the MRS staff.

F. (Shadi) Shahedipour-Sandvik
E. Fred Schubert
L. Douglas Bell
Vinayak Tilak
Andreas W. Bett

October 2009

MATERIALS RESEARCH SOCIETY SYMPOSIUM PROCEEDINGS

MATERIALS RESEARCH SOCIETY SYMPOSIUM PROCEEDINGS

Prior Materials Research Society Symposium Proceedings available by contacting Materials Research Society

Compound Semiconductors for
PV Applications

Mater. Res. Soc. Symp. Proc. Vol. 1167 © 2009 Materials Research Society   1167-O01-04

Utilizing Polarization Induced Band Bending for InGaN Solar Cell Design

Balakrishnam R Jampana[1], Ian T Ferguson[2], Robert L Opila[1], Christiana B Honsberg[3]
[1]Material Science and Engineering, University of Delaware, Newark, Delaware 19716, USA
[2]School of Electrical and Computer Engineering, Georgia Institute of Technology, Atlanta, Georgia 30332, USA
[3]Electrical Engineering, Arizona State University, Tempe, Arizona 85287, USA

ABSTRACT

Strong polarization effects observed in III-nitride materials can invert the surface carrier type. The corresponding band bending can be used to design InGaN solar cells. Similar surface inversion was observed in the past with silicon-based Schottky-barrier solar cells, but was limited by Fermi level pinning. The formation of two-dimensional electron gas by polarization fields in III-nitrides has been reported. Using a similar idea, the growth of a thin AlN capping layer on p-InGaN has resulted in band bending, hence depletion region, under the surface that can be used to separate any generated photo-carriers. Hall measurements at different depths on these structures confirm the inversion of surface carrier type. Solar cells based on this concept have resulted in an open circuit voltage of 2.15 V and short circuit current of 21.8 µA.

INTRODUCTION

Schottky-barrier solar cells invert the surface type of the semiconductors and are formed by metal deposition on semiconductors. The Schottky barrier height depends on the difference between metal work function and the semiconductor Fermi level. Silicon-based Schottky solar cells have been studied in the past [1]. The main limitation of these devices was Fermi level pinning to mid-band gap, limiting the open circuit voltage to half the band gap. Photo-response from III-nitrides Schottky devices were studied on GaN using several metal contacts. InGaN based Schottky devices have not yet been reported. Alternate approaches to invert the surface in other semiconductors have not been possible, but in III-nitride materials there is the inherent polarization which can be used to invert surfaces.

InGaN materials with their wide band gap range (0.7 to 3.4 eV) are ideal for solar cells to span most of the solar spectrum [2]. InGaN-based solar cells using quantum wells and p-i-n structures have been reported [3, 4]. A unique property of the III-nitrides is their inherent polarization property [5]. Strong polarization fields with formation of two-dimensional electron gas (2DEG) have been reported in III-nitrides and are used in high-electron mobility transistors [6]. The photo-response from surface inversion caused by the polarization property has not been explored in the past.

In this paper a study of surface inversion caused by thin AlN on InGaN is presented. The theoretically calculated band diagrams are presented followed by experimental evidence showing the formation of 2DEG and finally the photo-voltaic response from a p-InGaN/AlN structure. The III-nitride solar cells formed by polarization are not formed by inversion from a Schottky metal, however, these in principle demonstrate a similar behavior with an advantage of an optically transparent capping layer.

THEOR Y

Polarization is present in III-nitrides (0001) as a consequence of the non-centrosymmetry of the wurtzite structure and the large ionicity of the covalent metal-nitrogen bonds. The net polarization is composed of spontaneous and piezoelectric polarization [7]. The direction of piezoelectric polarization is dependent on the polarity of the material as well as on the strain. The spontaneous polarization in each epi-layer is calculated using the equations 1 below.

$$P_{sp}(In_xGa_{1-x}N) = -0.042x - 0.034(1-x) + 0.037\,x(1-x)\,(C/m^2) \qquad (1)$$

$$P_{sp}(AlN) = -0.09\,(C/m^2)$$

The piezoelectric polarization is dependent on the strain introduced by lattice-mismatch and is calculated using equation 2:

$$P_{pz} = 2\varepsilon_x \left[e_{31} - e_{33}\frac{C_{13}}{C_{33}} \right] \qquad (2)$$

Where, ε_x is strain at the InGaN-AlN interface, e_{31} (-0.58) and e_{33} (1.55) the elastic constants for AlN layer and C_{13} (115 GPa) and C_{33} (385 GPa) piezoelectric constant for AlN epi-layer.

The growth of pseudomorphic epi-layers results in formation of two-dimensional electron gas (2DEG) or two-dimensional hole gas at the interface, induced by polarization. 2DEG formation has been well studied, and using the above mentioned values as input into the "Silense" simulation program, the thermal equilibrium band diagrams are presented.

To study the extreme case AlN epi-layer on InGaN is selected. The band diagram of 100 nm p-InGaN with 5 nm AlN capping layer is shown in Figure 1. The formation of 2DEG can be clearly noticed in the band diagram. Solar cell operation requires a depletion region i.e. band bending. The formation of 2DEG in the III-nitrides introduces the required band bending for use as a solar cell and this is the study presented in this paper.

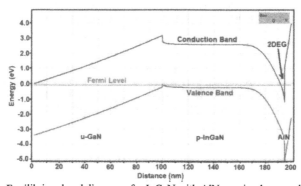

Figure 1. Equilibrium band diagram of p-InGaN with AlN capping layer under no bias

GROWTH and C HARA CTER IZ ATION
Ep ita xial Gro wth

The 100 nm thick p-InGaN layer with 5 nm AlN capping layer structure described above, and a 100 nm thick n-InGaN with 5 nm AlN capping layer, were grown on standard undoped-

GaN/sapphire templates in an Emcore MOCVD reactor. Trimethylgallium (TMG), trimethylaluminum (TMAl) and trimethylindium (TMI) were the precursors used to introduce gallium, aluminum and indium, respectively, into the reactor using hydrogen as the carrier gas; Ammonia (NH3) was used as the nitrogen source. Cp_2Mg was used for p-type doping and SiH_4 was used for n-type doping of InGaN. The u-GaN layer was grown using a two step process, where typically a 20-40 nm thin GaN buffer layer growth at 550°C was followed by the epitaxy of high quality 2 μm thick GaN template at 1030°C. The growth temperature for the InGaN layer was set to 740°C, and the AlN was grown at the same temperature to avoid the decomposition of InGaN.

Material and De vice Characterization

High-resolution X-ray Diffraction (HRXRD) studies were performed with an X'Pert diffractometer featuring Ge(220) hybrid four bounce monochromator in double-crystal mode with $CuK_{\alpha1}$ monochromatic radiation. The samples were characterized for optical reflection and transmission using PerkinElmer Lambda 1050 UV/Vis/NIR system in the 250 to 700 nm range. The p-InGaN was thermally treated at 650°C in N_2 ambient for 20 min to activate the p-type material. Time-integrated photoluminescence (PL) measurements on the InGaN samples were conducted in the temperature range of 8K to 120K. The excitation source was a 325 nm, 45 mW He-Cd continuous wave laser. Hall measurements were performed with an HMS-3000 system at room temperature using Van der Pauw arrangement. The Hall measurements were done on the surface of AlN and at a depth of 40 nm from surface after etching AlN epi-layer using inductively coupled plasma. The current-voltage (I-V) measurements were performed at room temperature under no illumination (dark) and AM 1.5 (1 sun) conditions using a custom built and calibrated system having an AM 1.5 filter from Newport-Oriel. The ohmic contacts were formed using In-Ga metal on the surface and at a depth of 40 nm from the surface.

RESULTS and DISCUSSI ON

The observed crystalline quality of the epitaxial layers is shown in Figure 2. The InGaN composition is determined to be around 15% for p-InGaN and 13.5% for n-InGaN layers. The fringes around the 15% p-InGaN layer indicate the absence of extended crystalline defects in the volume of InGaN layer that could affect the performance [9]. A broad AlN peak is also observed, and this thin layer is probably amorphous or poly crystalline due to the low growth temperature of AlN.

The PL at 8 K is shown in Figure 3(a) with multiple peaks around 3 eV. The band gap of the material is 3 eV and the peaks observed above 3 eV are related to the donor-acceptor pair transitions. The band gap is also confirmed to be 3 eV by the optical absorption shown in Figure 3(b). The optical absorption band edge, shown in Figure 3(b), is sharp, indicating absence of extended crystalline defects or phase separation in the InGaN epi-layers [10]. The solar cells studied here are wide band gap and hence the sharp band edge is required to transmit all the energy not absorbed by this active layer.

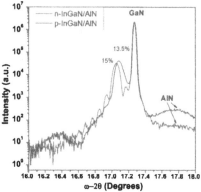

Figure 2. ω-2θ (0002) rocking curve for p-InGaN with AlN capping layer and n-InGaN with AlN capping layer

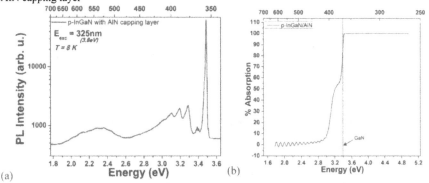

(a)

(b)

Figure 3. Optical properties of p-InGaN with AlN capping layers (a) Photo-luminescence at 8K (b) Optical absorption

The Hall measurement results for two locations shown in Figure 4 are tabulated in Table 1. The carrier concentration before thermal activation of p-InGaN measured on the AlN surface was $2.671×10^{13}$ cm^{-3}, while after activation it was measured to be $-2.021×10^{17}$ cm^{-3}, indicating formation of a 2DEG. Under similar conditions with n-InGaN epi-layer, the carrier concentration on the AlN surface was measured to be $-1.785×10^{19}$ cm^{-3}. The carrier concentrations below the AlN surface for p-InGaN and n-InGaN were measured to be $1.027×10^{18}$ and $-3.194×10^{18}$ cm^{-3}, respectively. Other parameters extracted from Hall measurement, mobility and resistivity, of epi-layers are also tabulated in Table 1. The mobility of p-InGaN and n-InGaN are in the range typically observed with epi-layers without any capping layer [11].

6

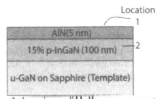

Figure 4. Location of Hall measurement

Table 1. Hall measurement results

Location / Condition	1		2
	Not Activated	Activated	Activated
p-InGaN /AlN capping layer			
Carrier Concentration (cm⁻³)	2.671×10^{13}	-2.021×10^{17}	1.027×10^{18}
Mobility (cm²/Vs)	2.942	0.6109	5.978
Resistivity (Ω-cm)	7.944×10^{4}	50.55	1.017
n-InGaN /AlN capping layer			
Carrier Concentration (cm⁻³)	-	-1.785×10^{19}	-3.194×10^{18}
Mobility (cm²/Vs)	-	54.49	18.83
Resistivity (Ω-cm)	-	6.418×10^{-3}	0.1038

The photo-response of the device is shown in Figure 5. The open circuit voltage of 2.15 V and short circuit current of 21.8 µA are observed. A very good diode action is observed under no illumination, but with illumination under AM 1.5 spectrum, a change in series and shunt resistance is observed with fill factor less than 5.

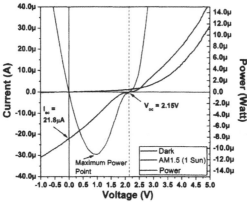

Figure 5. Current-Voltage characteristics under dark and AM1.5 (1 sun) conditions

Conventional Schottky-barrier solar cells are formed by surface inversion using Schottky metal contacts. The limitation with these solar cells lies with the aerial coverage originating from shadowing by metal contacts. The depletion region width in Schottky solar cell structures is formed below and in short range around the Schottky metal contact. The polarization in III-nitrides can be used to invert the surfaces, and this inversion is caused by optically transparent epi-layers. A few drawbacks of the polarization based structures are formation of a potential well at the InGaN/AlN interface and the optimum thickness of AlN epi-layer. The optimum thickness of AlN epi-layer should be thick enough to cause strong band bending and thin enough for photo-carrier to tunnel.

CONCLUSIONS

Schottky-barrier type InGaN solar cells using inherent polarization can be used to design III-nitride solar cells. The simulated band diagrams and the supporting experimental device structures were presented. The band diagrams indicate a clear 2DEG formation and the corresponding band bending can be used to design solar cells. The Hall measurements indicate the surface type inversion and 2DEG formation with AlN capping layer on p-InGaN. The photovoltaic response from the surface inverted device structure was demonstrated, with 2.15 V open circuit voltage and 21.8 μA short circuit current. Although the fill factors are low, these can be improved with improvements in ohmic contacts and by addressing the current spreading issues commonly observed in InGaN solar cells.

REFERE NCES

1. M. A. Green, Journal of Applied Physics, vol. 47, pp. 547-54, 1976.
2. O. Jani, I. Ferguson, C. Honsberg, and S. Kurtz, Applied Physics Letters, vol. 91, p. 132117, 2007
3. R. Dahal, B. Pantha, J. Li, J. Y. Lin, H. X. Jiang, Applied Physics Letters, v 94, n 6, p 063505, Feb. 2009
4. C.J. Neufeld, N.G. Toledo, S.C. Cruz, M. Iza, S.P. DenBaars, U.K. Mishra, Applied Physics Letters, v 93, n 14, p 143502, Oct. 2008
5. O. Ambacher, J. Smart, J. R. Shealy, N. G. Weimann, K. Chu, M. Murphy,W. J. Schaff, L. F. Eastman, R. Dimitrov, L. Wittmer, M. Stutzmann, W.Rieger, and J. Hilsnbeck, J. Appl. Phys. vol. 85, p. 3222, 1999
6. A. E. Romanov, T. J. Baker, S. Nakamura et al., Journal of Applied Physics 100 (2), 023522 (2006)
7. V. Fiorentini F. Bernardini, Phys. Stat. Sol. B, 216 (1999), p. 391
8. J. Osvald, Applied Physics A: Materials Science and Processing 87 (4), 679-682 (2007)
9. Nikolai Faleev, Balakrishnam Jampana, Anup Pancholi, Omkar Jani, Hongbo Yu, Ian Ferguson, Valeria Stoleru, Robert Opila, and Christiana Honsberg, Proceedings of 33rd IEEE PVSC, San Diego, May 2008
10. Balakrishnam R Jampana, Nikolai N Faleev, Ian T Ferguson, Robert L Opila, Christiana B Honsberg, "Crystalline Perfection of Epitaxial Structure: Correlation with Composition, Thickness, and Elastic Strain of Epitaxial Layers, MRS Spring 2009, San Francisco, April 2009
11. Hongbo Yu, Andrew Melton, Omkar Jani, Balakrishnam Jampana, Shenjie Wang, Shalini Gupta, John Buchanan, William Fenwick, and Ian Ferguson, "MOCVD Growth of High-Hole Concentration (>2×10^{19} cm^{-3}) p-type InGaN for Solar Cell Application", MRS 2008 Fall Meeting, Boston, USA, Nov. 2008

Compound Semiconductors
for Lighting

Mater. Res. Soc. Symp. Proc. Vol. 1167 © 2009 Materials Research Society 1167-O02-05

Enhancement of InGaN-based MQW Grown on Si(111) Substrate by Underlying AlGaN/GaN SLS Cladding Layer

Bin Abu Bakar Ahmad Shuhaimi[1,2,*], Takaaki Suzue[1,2], Yukiyasu Nomura[1,2], Yoshinori Maki[1,2] and Takashi Egawa[1,2]

[1]Research Center for Nano-Device and System, Nagoya Institute of Technology, Nagoya, Japan.
[2]Graduate School of Engineering Physics, Electronics and Mechanics, Nagoya Institute of Technology, Nagoya, Japan.

ABSTRACT

This paper reports enhanced internal-quantum-efficiency (IQE) in InGaN-based multi-quantum-well (MQW) grown on Si(111) substrate with underlying strained-layer-superlattice (SLS) cladding layer for application in LDs and LEDs. In comparative study between a thick $Al_{0.03}Ga_{0.97}N$ bulk and an $Al_{0.06}Ga_{0.94}N$/GaN SLS cladding layer, transmission-electron-microscopy (TEM) images reveal that $Al_{0.06}Ga_{0.94}N$/GaN SLS cladding layer is effective to suppress threading dislocations. A higher IQE has been achieved in sample with underlying $Al_{0.06}Ga_{0.94}N$/GaN SLS cladding layer, compared to that of $Al_{0.03}Ga_{0.97}N$ bulk cladding layer. IQE of 31.6% has been achieved in sample with underlying $Al_{0.06}Ga_{0.94}N$/GaN SLS cladding layer when the MQW thickness is reduced to 2 nm.

INTRODUCTION

GaN-based wide band-gap emitters grown on Si are very promising breakthrough for low-cost energy-saving applications. Silicon (Si) is a very promising substrate for GaN growth, which is low cost, widely available, having large surface diameter, and offering good current and thermal conductivity for device operation. However, a few issues that hindered epitaxial-growth of GaN on Si(111) substrate are cloudy and cracked epilayer, large wafer curvature and high threading dislocation density exceeding 10^{10} cm^{-2} in conventional metalorganic chemical vapor deposition (MOCVD) method. These are attributed by large lattice and thermal mismatch between GaN epilayer and Si(111) substrate of 17% and 116%, respectively. A thin AlN layer is generally used as seeding layer for GaN growth on Si substrate, followed by methods such as AlN/GaN or AlGaN/GaN multilayer [1-5], AlN interlayer [6-7], Si_xN_y interlayer [8-9], etc, to offset tensile and thermal stress between GaN and Si, and to improve crystal quality in the subsequent epilayer.

Despite the difficulties for GaN growth on Si, a few groups including ours have succeeded in growing good epitaxial quality GaN layer on Si(111) substrate with applications in LEDs [2-5] and HEMTs [10]. In this paper, we report an efficient suppression of threading dislocations in InGaN-based MQW sample grown on Si(111) substrate by an underlying $Al_{0.06}Ga_{0.94}N$/GaN strained-layer-superlattice (SLS) cladding layer, and an improvement of MQW internal-quantum-efficiency (IQE) by optimizing MQW thickness in sample with underlying $Al_{0.06}Ga_{0.94}N$/GaN SLS cladding layer. This work is intended for applications in GaN-based LDs and LEDs on Si substrate.

---

*Corresponding author, e-mail: shuhaimi@msn.com

EXPERIMENT

Samples in this study were grown on 2 inch Si(111) substrate by horizontal-reactor MOCVD. Trimethylgallium (TMGa), trimethylaluminum (TMAl) and ammonia (NH$_3$) were used as precursors for Ga, Al and N, respectively. Hydrogen (H$_2$) was used as carrier gas. Monosilane (SiH$_4$) diluted in hydrogen was used for n-type dopant. Prior to growth, the substrate was thermally cleaned at 1100 °C in H$_2$ flow. A 20 nm thin AlN layer was grown on the Si(111) substrate as seeding layer, followed by 40 pairs of AlN/GaN multilayers (MLs) with respective thickness of 5 nm and 20 nm. Two types of cladding layer were grown on the MLs, (a) with a thick Al$_{0.03}$Ga$_{0.97}$N bulk layer, and (b) with SLS layer consisting of Al$_{0.06}$Ga$_{0.94}$N/GaN pairs. Subsequently, a 60 nm GaN waveguide layer and an active layer, with 3 pairs of 4 nm In$_{0.16}$Ga$_{0.84}$N quantum well and 9 nm of Si-doped In$_{0.08}$Ga$_{0.92}$N barrier were grown. Finally, the growth was capped with a 10 nm GaN layer. Furthermore, optical characteristics in sample with underlying Al$_{0.06}$Ga$_{0.94}$N/GaN SLS cladding layer was optimized by reducing the MQW thickness to 3 nm and 2 nm. MQW pairs and other growth parameters were kept similar in all samples.

Structural properties were evaluated by transmission electron microscopy (TEM), observed using JEOL JEM-2010F FasTEM system operating at 200 kV. TEM samples were prepared by a standard procedure, with thickness of 100 nm. Optical characteristics were evaluated by temperature-dependence photoluminescence (PL) method, using a closed-circuit helium cryostat equipped with heater and temperature controller. A 325 nm He-Cd laser was used for PL excitation.

DISCUSSION

All samples in this study are crack-free, having specular surface and free from melt-back etching. TEM cross-sectional bright-field images of the samples are shown in figure 1, with sample of Al$_{0.03}$Ga$_{0.97}$N bulk cladding layer shown in (a) and (b), and sample of Al$_{0.06}$Ga$_{0.94}$N/GaN SLS cladding layer shown in (c) and (d). Major layers are marked in the figures. The TEM images are taken with $g = [0002]$ and $g = [11\bar{2}0]$ diffraction conditions, making screw dislocation with $b = [0001]$ and edge dislocation with $b = (1/3)[11\bar{2}0]$ visible in respective diffraction. Mixed dislocation with $b = (1/3)[11\bar{2}3]$ can be observed in both planes, according to the $g \times b$ invisibility criterion.

A high density of threading-dislocations (TDs) consisting of stacking faults, fault loops and coalescing dislocations can be observed originating at the AlN seeding layer, which is attributed by large lattice and thermal mismatch with the Si(111) substrate. However, TDs are bent and reduced as the AlN/GaN ML pair is increased. TDs which penetrate to the upper edge of the MLs surpass the interface and further penetrate into the cladding layer. In sample with thick Al$_{0.03}$Ga$_{0.97}$N bulk cladding layer, formation of new dislocations can also be observed at the interface due to sudden change of lattice constant between AlN/GaN MLs and Al$_{0.03}$Ga$_{0.97}$N cladding layer. TDs at the lower edge in the Al$_{0.03}$Ga$_{0.97}$N cladding layer bend, and a few TDs intersect into a single TD propagating vertically towards sample surface which is normally seen in a thick relaxed GaN-based layer.

In contrast to that phenomenon, in sample with Al$_{0.06}$Ga$_{0.94}$N/GaN SLS cladding layer, TDs surpassing the interface between AlN/GaN MLs and Al$_{0.06}$Ga$_{0.94}$N/GaN radically disappear at the lower region in the $Al_{0.06}Ga_{0.94}N$/GaN SLS cladding layer. Interestingly, we do not observe any formation of new TD at the interface. TDs that surpass the interface bend abruptly and annihilate in the $Al_{0.06}Ga_{0.94}N$/GaN SLS cladding layer. The initial pair of $Al_{0.06}Ga_{0.94}N$/GaN SLS cladding layer is compressive to resemble lattice constants of the underlying AlN/GaN MLs. This behavior can prevent formation of new TD at the interface. The SLS gradually returns to its stable strain state as the SLS pair number is increased. The existence of strain between $Al_{0.06}Ga_{0.04}N$ and GaN layer in the SLS pair forces TDs to bend at abrupt angle and annihilate in the cladding layer.

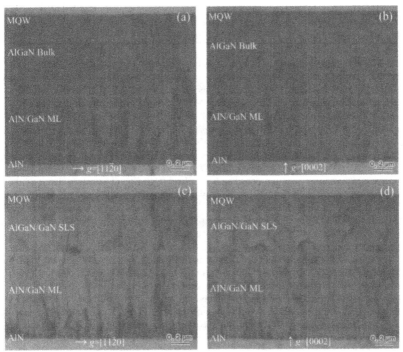

Figure 1. TEM cross-sectional bright-field images of epitaxial structure on Si(111) substrate in this study. Sample with $Al_{0.03}Ga_{0.97}N$ bulk cladding layer is shown in (a) and (b), while sample with $Al_{0.06}Ga_{0.94}N$/GaN SLS cladding layer is shown in (c) and (d). Diffraction plane is marked in each figure.

Threading dislocation density (TDD) in the upper region of each cladding layer is summarized in Table I. The TDD estimation was determined by counting along crystal plane normal to the growth direction from TEM micrograph of 5 μm length and 100 nm sample thickness. Edge and mixed type of dislocations are predominant in both samples. We find that TDD of screw and mixed type at $g = [0002]$, and edge and mixed type at $g = [11\bar{2}0]$ in sample with $Al_{0.06}Ga_{0.94}N$/GaN SLS cladding layer is reduced compared to that of sample with thick $Al_{0.03}Ga_{0.97}N$ bulk cladding layer. This confirms that SLS cladding layer is effective to annihilate threading dislocations in GaN-based heteroepitaxy on Si(111) substrate.

Table I. Threading dislocation density in the upper region of each type of cladding layer determined from TEM micrograph.

Type of Cladding Layer	Threading Dislocation Density (cm^{-2})	
	Screw and Mixed $g = [0002]$	Edge and Mixed $g = [11\bar{2}0]$
$Al_{0.03}Ga_{0.97}N$ Bulk	8.4×10^9	1.5×10^{10}
$Al_{0.06}Ga_{0.94}N$/GaN SLS	3.4×10^9	6.5×10^9

MQW optical performance in each underlying cladding layer was evaluated by temperature-dependence PL. Figure 2 shows Arrhenius plot of normalized integrated intensity for the MQW emission with 10 K temperature increment between 10 K and 100 K, and 20 K increment subsequently until 300 K. IQE of the MQW can be estimated by comparing PL normalized integrated intensity ratio between 300 K and 10 K [11]. Assuming that the IQE at 10 K is 100%, estimation of IQE at 300 K for sample with identical 4 nm MQW thickness is 2.1% and 4.8% in sample with underlying $Al_{0.03}Ga_{0.97}N$ bulk and $Al_{0.06}Ga_{0.94}N$/GaN SLS, respectively. The IQE enhancement is attributed by efficient suppression of TDs by the SLS cladding layer, as discussed earlier. Further improvement of the IQE can be achieved by reducing the MQW layer thickness, which reduces quantum-confined Stark effect (QCSE) and increases electron-hole wave-function overlap in the InGaN-based MQWs [12]. Sample with 3 and 2 nm MQW thickness shows respective estimated IQE of 11.2% and 31.6%, when grown with underlying $Al_{0.06}Ga_{0.94}N$/GaN SLS cladding layer. The steep decrease of integrated intensity in temperature range between 120 K and 160 K is attributed by activation of non-radiative process in the recombination [13].

Figure 3 shows temperature dependence normalized PL intensity for sample with 2 nm thickness of $In_xGa_{1-x}N$ MQW grown with underlying $Al_{0.06}Ga_{0.94}N$/GaN SLS cladding layer. The emission at 2.9 eV originates from radiative recombination of excitons in the wells, while emission at 3.1 eV which is visible at low temperature originates from $In_yGa_{1-y}N$ barriers. The intensity ratio at 300 K and 10 K is consistent with the estimated IQE in figure 2. These results confirm that $Al_{0.06}Ga_{0.94}N$/GaN SLS cladding layer is effective to suppress TDs and therefore enhance optical emission in InGaN-based MQW grown on Si(111) substrate with a relatively low pair of MQWs.

CONCLUSIONS

In conclusion, we have reported reduction of threading dislocations for GaN growth on Si(111) by an underlying $Al_{0.06}Ga_{0.94}N$/GaN SLS cladding layer. TEM images confirm that the SLS cladding layer efficiently suppresses TDs from propagating into subsequently grown layer. A relatively high IQE of 31.6% has been achieved in sample with the underlying SLS cladding layer, when MQW thickness is optimized to 2 nm.

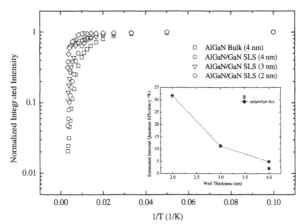

Figure 2. Temperature dependent normalized PL integrated intensity for Al0.03Ga0.97N bulk cladding layer and Al0.06Ga0.94N/GaN SLS cladding layer. Inset: Estimated PL IQE as a function of well thickness.

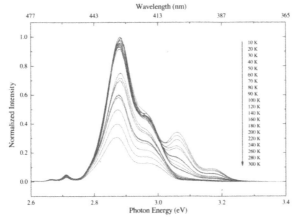

Figure 3. Temperature dependent normalized PL intensity for 3 pair of 2 nm thickness InGaN-based MQWs grown with underlying $Al_{0.06}Ga_{0.94}N$/GaN SLS cladding layer on Si(111) substrate.

REFERENCES

1. H. Ishikawa, G. Y. Zhao, N. Nakada, T. Egawa, T. Jimbo and M. Umeno, Jpn. J. Appl. Phys. **38**, L492 (1999).
2. T. Egawa, B. Zhang, N. Nishikawa, H. Ishikawa, T. Jimbo and M. Umeno, J. Appl. Phys. **91**, 528 (2002).

3. T. Egawa, T. Moku, H. Ishikawa, K. Ohtsuka and T. Jimbo, Jpn. J. Appl. Phys. **41**, L663 (2002).
4. T. Egawa, B. Zhang and H. Ishikawa, IEEE Elect. Dev. Lett. **26**, 169 (2005).
5. B. Zhang, T. Egawa, H. Ishikawa, Y. Liu and T. Jimbo, Jpn. J. Appl. Phys. **42**, L226 (2003).
6. A. Dadgar, J. Christen, T. Riemann, S. Richter, J. Bläsing, A. Diez, A. Krost, A. Alam and M. Heuken, Appl. Phys. Lett. **78**, 2211 (2001).
7. A. Dadgar, C. Hums, A. Diez, J. Bläsing and A. Krost, J. Cryst. Growth **297**, 279 (2006).
8. K. Cheng, M. Leys, S. Degroote, M. Germain and G. Borghs, Appl. Phys. Lett. **92**, 192111 (2008).
9. K. Y. Zang, Y. D. Wang, L. S. Wang, S. Y. Chow and S. J. Chua, J. Appl. Phys. **101**, 093502 (2007).
10. S. L. Selvaraj, T. Ito, Y. Terada and T. Egawa, Appl. Phys. Lett. **90**, 173506 (2007).
11. J. K. Son, S. N. Lee, T. Sakong, H. S. Paek, O. Nam, Y. Park, J. S. Hwang, J. Y. Kim and Y. H. Cho, J. Cryst. Growth 287, 558 (2006).
12. S. F. Chichibu, A. Shikanai, T. Deguchi, A. Setoguchi, R. Nakai, E. Nakanishi, K. Wada, S. P. DenBaars, T. Sota and S. Nakamura, Jpn. J. Appl. Phys. **39**, 2417 (2000).
13. Y. H. Cho, G. H. Gainer, A. J. Fischer, J. J. Song, S. Keller, U. K. Mishra and S. P. DenBaars, Appl. Phys. Lett. **73**, 1370 (1998).

Poster Session:
Compound Semiconductors for PV

Mater. Res. Soc. Symp. Proc. Vol. 1167 © 2009 Materials Research Society   1167-O03-04

InGaP Layers Grown on Different GaAs Surfaces for High Efficiency Solar Cells

O. Martínez[1], V. Hortelano[1], V. Parra[2], J. Jiménez[1], T. Prutskij[3], C. Pelosi[4]

[1] GdS-Optronlab, Dpto. Física de la Materia Condensada, Edificio I+D, Paseo de Belén 1, 47011 Valladolid, Spain
[2] Grupo Pevafersa, Av. Vicente Fernández Manso, 49800 Toro (Zamora), Spain
[3] Instituto de Ciencias, BUAP, Apartado Postal 207, 72000 Puebla, México
[4] IMEM-CNR, viale Usberti 37/A, 43100 Parma, Italy

ABSTRACT

InGaP layers grown on (111)Ga and (111)As GaAs substrate faces are investigated by microRaman spectroscopy, microphotoluminescence and cathodoluminescence. The growth on these polar faces benefits disorder with respect to the layers grown on (001) faces. It is shown that both (111)Ga and (111)As faces result in disordered InGaP layers. While the layers grown on (111)As faces present inhomogeneous compositions, the layers grown on (111)Ga faces present homogeneous compositions close to lattice matching and are almost disordered.

INTRODUCTION

Nowadays, there is a renewed interest on high efficiency solar cells, based on multi-junction structures of III-V compounds [1]. The ternary alloy InGaP is an essential material for multijunction cells. InGaP lattice matched to GaAs presents very interesting properties, as its direct bandgap of 1.9 eV. On the other hand, it presents some advantages respect to AlGaAs, as the absence of DX centers - which are responsible for a drastic reduction of the free carrier concentration in AlGaAs [2] -, larger band gap, and a more favourable band gap alignment with GaAs. It has been suggested that it might be useful for solar cells to make use of the growth on (111) GaAs faces under mismatched conditions, where large internal electric fields are generated by the off-diagonal strain [3]. In fact, the electronic band structure of QWs on (111) faces is substantially modified. The piezoelectric fields could allow for a better extraction of minority carriers and for a larger absorption coefficient [4]. Also, one might expect suppression of long range order, which is responsible for the band gap shrinkage and the change of the band alignment between InGaP and GaAs [5]; the presence of order has adverse consequences for the device performance. The growth modes on (111) faces are different from those on (001) faces, because of the different dangling bonds exposed and the different surface reconstructions. The optimization of the InGaP layers grown on (111) faces could provide a useful route to achieve better photovoltaic devices. This optimization demands a significant characterization effort, aiming to study the main structural and optical properties of the InGaP layers grown on (111) faces. This characterization needs to address problems as the control of the composition (short range order), spontaneous ordering (long range order), and the presence of defects. We explore here the preliminary growth by metal-organic chemical vapour deposition (MOCVD) of InGaP layers on (111)Ga and (111)As - GaAs substrates, and compare the results with the layers grown in the same conditions on (001) GaAs substrates.

Changes in the growth rate and composition are expected depending on the surface orientation used, because of the differences on the surface diffusion and sticking of the constituent atoms on the different surfaces. In order to control these changes, as well as the presence of ordered domains, optical measurements, microRaman (μR), microphotoluminescence (μPL), and cathodoluminescence (CL) were carried out. In that way, the optical properties of the different layers have been studied. The problems of ordering, phase separation and homogeneity are addressed, showing that InGaP/(111)Ga - GaAs structures present interesting perspectives for growing homogeneous disordered InGaP material.

EXPERIMENT

The layers were grown at 600°C under a reduced pressure of 60 mbar using TMGa and TMIn, arsine and phosphine as main reagents. The MOVCD apparatus is a horizontal shaped one. The heating system consisted of a battery of 6 IR lamps focused to a graphite susceptor. In order to study the relationship between polarity and growth conditions, three GaAs substrates, (001), (111)Ga, and (111)As were used to growing simultaneously InGaP layers, with the lattice matched composition ($In_{0.49}Ga_{0.51}P$). Three different flows of phosphine, namely 290, 320 and 350 cm^3/min were used, keeping all other growth conditions unchanged, in order to study the effect of the V/III ratio on the layer properties. The growth on (111) faces, either Ga or As terminated, drastically changes the reagent adsorption and surface reaction mechanisms, which affects the growth rates and the distribution of the atomic species; therefore, changes in the layer composition and the growth rate might take place.

μPL measurements were carried out at room temperature exciting with an Ar^+ laser (514.5 nm), in a Labram UV-HR800 spectrometer (Jobin Yvon). The Raman spectra were measured in the same apparatus; therefore the μR and μPL data are fully comparable, since they were acquired in the same points of the samples.

CL measurements were carried out with a Xiclone –CL sytem from Gatan. The measurements were carried out at 80 K. The samples grown on (111) facets gave a much poorer CL signal than the samples grown on the (001) facet.

RESULTS AND DISCUSSION

The Raman spectrum of InGaP presents a two-mode behaviour [6]. It consists of a LO phonon mode (henceforth labelled LO_1) (GaP like) peaking at 381 cm^{-1} for GaAs lattice matched InGaP; a TO phonon mode (InP-like, TO_2) at 330 cm^{-1}, which is selection rule forbidden, although it is activated by alloy disorder; a phonon band appearing at 365 cm^{-1} has been associated with the InP-like LO phonon (LO_2) [7]. The frequencies and relative intensities of the modes are dependent on the composition, while the crystal orientation influences the relative intensity of the different phonon modes. Typical Raman spectra are shown in Fig.1a. One observes the LO_1 and LO_2 modes as the dominant ones in samples grown on (001) substrates. In samples grown on (111) substrates the TO_2 mode is strongly enhanced because of the crystal orientation. The Raman spectra of samples grown on (001) and (111)Ga substrates present a very good homogeneity over the full surface; however the Raman spectra of samples grown on (111)As substrates are highly inhomogeneous over the wafer (the spectra at two different points are shown).

The Raman shift of the LO_1 phonon peak has been claimed not to be affected by the spontaneous order [8]; therefore, it can be used to give an estimation of the $In_{1-x}Ga_xP$ composition according to the following relation:

$$\omega_{LO1}=346.84 + 76.33 \text{ x} - 18.18 \text{ x}^2 \qquad (1)$$

The data obtained for the different samples are reported in table I, where the corresponding composition is calculated with eq.(1). One observes that the samples grown on (001) substrates are very close to the lattice matched composition; some deviation and inhomogeneity were observed for the lowest phosphine flow. The samples grown on (111)Ga substrates are also close to lattice matching, but they present in average a slightly lower x. Finally, for samples grown on (111)As substrates the LO_1 phonon frequency presents a large dispersion, being generally shifted to the low frequencies, which accounts for In-rich compositions. In general, it seems that low phosphine flow benefits the incorporation of In, specially in the case of (111)' faces. These data were checked over a large number of points, so they really reflect the properties of the different samples.

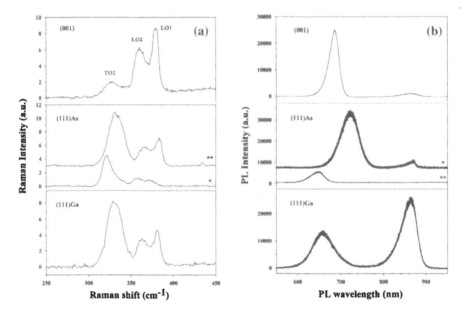

Figure 1. a) Raman and b) PL spectra of the studied samples. (Two points are shown for sample (111)As)

Table I. List of the studied samples, phosphine flux, LO$_1$ Raman peak position, estimated x from Raman shift, Eg values: Eg calc. (from the x values), Eg exp. (from the PL peak), and ΔEg (=Eg calc-Eg exp) (see text)

Face	Phosphine (cm^3/min)	ω_{LO1} (cm^{-1})	x	Eg calc. (eV)	Eg exp. (eV)	ΔEg (eV)
(001)	290	378.4–387.1	0.465–0.619	1.84–2.07	1.77–2.00	0.07
	320	381.2	0.513	1.91	1.80	0.11
	350	381.2	0.513	1.91	1.81	0.10
(111) As	290	364.3–383.5	0.243–0.553	1.57–1.97	1.59–1.94	-0.02–0.03
	320	373–383.4	0.376–0.551	1.72–1.97	1.71–1.97	0–0.01
	350	370.2–381.9	0.332–0.525	1.67–1.93	1.64–1.93	0–0.03
(111) Ga	290	379.6	0.485	1.87	1.83	0.04
	320	381.4	0.516	1.91	1.89	0.02
	350	379.5	0.484	1.87	1.87	0

The PL spectrum was acquired in the same points where the Raman spectra were taken. Typical PL spectra are shown in Fig.1b. For samples grown on (001) substrates one observes the InGaP emission, which the peak energy fluctuates around 690 nm (1.8 eV), without noticeable changes in the Raman spectra and the emission arising from the substrate. For samples grown on (111)Ga substrates the spectra are similar but the intensities are reversed, being the InGaP emission very weak compared to the (001) substrate, which is due to the much thinner InGaP layer on (111)Ga due to the lower growth rate on this face; the InGaP emission peaks around 667 nm (1.86 eV). Finally, the spectra of layers grown on (111)As substrates present the InGaP peak with a large spatial fluctuation, in agreement with the compositional changes reported from Raman data; on the other hand, some points exhibit a double peak, showing phase separation, with In-rich regions.

The E$_g$ values estimated from the InGaP peak energy are reported in table I (E$_g$ exp.). An estimation of the E$_g$ values corresponding to the composition deduced from the LO$_1$ frequency was also done (E$_g$ calc). The band gap energy at room temperature of In$_{1-x}$Ga$_x$P ternary alloy free of strain and spontaneous order obeys to the following expression [9]:

$$E_g = 1.35 + 0.73x + 0.7x^2 \qquad (2)$$

The values calculated are also reported in table I (E$_g$ calc.), as well as the differences with the values obtained from the room temperature PL peak, ΔE$_g$.

One observes a very good matching between the two E$_g$ values, experimental and calculated, for samples grown on (111)As substrates, which suggests the absence of order on these samples. The samples grown on the other face, (111)Ga, also give E$_g$ values close to the value corresponding to their composition, which points to very low order degree. Finally the largest ΔE$_g$, and therefore the largest degree of order, are observed for the samples grown on (001) substrates.

CL images are shown in Fig.2 for the samples grown on (001) and (111)As faces. Fig.2a shows the peak wavelength distribution of the InGaP layer grown on the (001) face. One observes a cross hatched distribution, with an overall variation of the peak wavelength of ≈5 nm. The CL intensity of the InGaP emission also evidences the same cross hatched distribution, Fig.2b, which would be due to the charge confinement in the low energy domains. The images of the substrate emission also evidence the cross hatched distribution. The peak wavelength distribution of the substrate emission, Fig.2c, is very similar to the distribution of the peak wavelength of the InGaP layer. However, they are anticorrelated, and the overall wavelength variation is only ≈1 nm, which is much smaller that the peak wavelength fluctuation in the InGaP layer; this suggests that the peak wavelength shifts of the InGaP layer are due to ordering, rather than other effects as strain. The intensity of the substrate emission also follows the same pattern, Fig.2d, although the cross hatched structure is less defined. The distributions observed in the substrate are probably due to Si doping fluctuations. The n-type doping of the substrate is known to influence the degree of order [10]. These images evidence the relevant role of the substrate on the formation of the ordered domains.

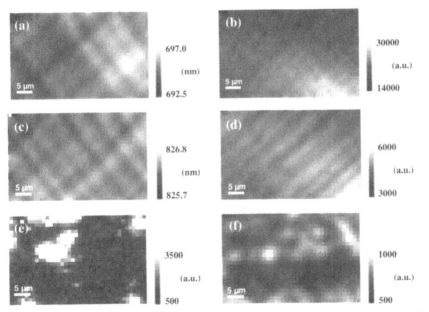

Fig.2. a) Peak wavelength distribution of the InGaP emission, b) monochromatic image at ~694 nm (corresponding to the InGaP emission), c) peak wavelength distribution of the substrate emission and d) monochromatic image at ~826 nm (corresponding to the GaAs emission) of a sample grown on a (001) face; e) monochromatic image at 656 nm (wavelength corresponding to nearly lattice matched composition) and f) monochromatic image at 710 nm (wavelength corresponding to In-rich regions), of a layer grown on a (111)As face.

The samples grown on (111)As substrates present islands with compositional variations, Fig. 2e) and f), which is in full agreement with the PL and Raman data that show a large compositional inhomogeneity of these samples. These islands can be propitiated by the very different diffusion lengths of In and Ga on this surface [11]. Finally, the samples grown on (111)Ga substrates appear more homogeneous than the other samples. However, the low growth rate difficult to obtain CL images because of the thin InGaP layers.

CONCLUSIONS

Ordering is substantially reduced in (111)As and (111)Ga faces; in particular, it is almost inexistent in (111)As faces, and very small in (111)Ga faces. Nevertheless, other problems have to be solved to grow high quality InGaP layers on the (111) faces, i.e. the control of the composition, which is particularly critical for samples grown on (111)As substrates, and the inhomogeneous growth. The layers grown on (111)Ga substrates present a low growth rate, and a tendency to be slightly In rich, in spite of the low degree of order and the good homogeneity. This preliminary research suggests that disordered InGaP can be grown on (111) GaAs substrates. Further growth runs modifying the growth parameters might presumably allow the improvement of the InGaP layers.

ACKNOWLEDGMENTS

The Spanish group of authors acknowledges the financial support from Junta de Castilla y León (GE-202)

REFERENCES

1. M. Yamaguchi, T. Takamoto, and K. Araki, Solar Energy Materials & Solar Cells 90, 3068 (2006).
2. P.M. Mooney, J. Appl. Phys. 67, R1 (1990).
3. D.L. Smith, Sol. St. Commun. 57, 919 (1986).
4. M. Hopkinson, J.P.R. David, E.A. Khoo, A.S. Pabla, J. Woodhead, and G.J. Rees, J. Microelectron. 26, 805 (1995).
5. Y.Zhang, A.Mascarenhas, and L.W. Wang, Appl. Phys. Lett. 80, 3111 (2002).
6. S. H. Wei and A. Zunger, Phys. Rev. B 49, 14337 (1994).
7. G. Lucovsky, M.H. Brodsky, M.F. Chen, J. Chicotka, and A.T. Ward, Phys. Rev. B 4, 1945 (1971).
8. M. Zachau, and W.T. Masselink, Appl. Phys. Lett. 60, 2098 (1992).
9. G. B. Stringfellow, J. Appl. Phys. 43, 3455 (1972).
10. S. Scardova, C. Pelosi, G. Attolini, B. Lo, O. Martínez, E. Martín, A. M. Ardila, and J. Jiménez, Phys. Stat. Sol. (a) 195, 50 (2003).
11. M.M.G. Bongers, P.L. Bastos, M.J. Anders, and L.J. Giling, Journal of Crystal Growth 171, 333 (1997).

Mater. Res. Soc. Symp. Proc. Vol. 1167 © 2009 Materials Research Society   1167-O03-11

Fabrication of TiO2 Nanobelt Network for Dye-Sensitized Solar Cells

Haiyan Li and Jun Jiao*
Department of Physics, Portland State University,
Portland, OR 97207, U.S.A.
*Corresponding author. E-mail: jiaoj@pdx.edu

ABSTRACT

Interconnected TiO$_2$ nanobelt networks were prepared to serve as anode materials. The aim is to enhance the electron transport through the anode of dye-sensitized solar cells. Using an alkaline hydrothermal procedure and by controlling the reaction time two kinds of nanostructures were synthesized. One is TiO$_2$ nanobelts and another is TiO$_2$ nanobelts protruding from TiO$_2$ nanoparticles. The investigation suggests that TiO$_2$ nanobelts resulted from the rearrangement of the adjacent $[Ti(OH)_6]^{2-}$ monomers formed during the erosion process of TiO$_2$ nanoparticles. The nanostructures of as-synthesized nanobelts were woven and interconnected, forming networks after an annealing process. Raman analysis indicated that both kinds of nanostructures were pure anatase. Electrical characterization suggests that the conductivities of these TiO$_2$ nanobelt networks were higher than those of the TiO$_2$ nanoparticle films. Under simulated sunlight with an intensity of AM 1.5 G, the solar cells made of TiO$_2$ nanobelt networks show exceptional photocurrent in comparison to those made of TiO$_2$ nanoparticles.

INTRODUCTION

The photosensitization of wide band gap semiconductor, such as TiO$_2$ (3.4 eV), is a promising and low cost method (reel-to-reel mode for mass production) to fabricate dye sensitized solar cell (DSSC), in which the sensitizer generates electron-hole pairs after harvesting photons and then injects the excited electrons into TiO$_2$ conduction band. Subsequently, the electrons and holes are collected by the opposing electrodes correspondingly.[1-3] Using a ruthenium complex sensitized TiO$_2$ nanoparticle (NP) based photoanode, the DSSC is capable of absorbing most of visible sunlight and a high light-electricity power conversion efficiency (η_{PCE}) of 11 % was obtained.[4] After photogenerated excitons were split, the efficiency of charge collection is related to the competition between charge transport and electron-hole recombination.[5-6] Since the diffusion of oxidizing species is exceedingly slow, the enhancement of electron transport will reduce the charge recombination. Because of many surface defects and crystal boundaries in the TiO$_2$ NP layer, the electron transport from the excited dye molecules to the electrode is slow and follows a trap-mediated diffusion mechanism.[7] Using TiO$_2$ nanorods (NR), nanowires (NW) or nanobelts (NB) instead of TiO$_2$ NPs can significantly increase the electron diffusion length up to over tens of micrometers by removing the boundaries between TiO$_2$ NPs.[8, 9] ZnO NWs (wide band gap of 3.3 eV, similar to that of TiO$_2$) are also used as the dye carriers and electron transport materials showing one or two order of magnitude enhancement of the electron transport relative to that of NP based cell.[10] Nevertheless, the reported one dimensional materials based DSSCs have very low conversion efficiency, because the replacement of small size NPs by NRs, NWs or NBs leads to a pronounced loss of specific surface area. Therefore it is desirable to prepare a materials having high surface area similar to that of the TiO_2 NPs and long electron diffusion length.

In this work, we synthesized the TiO_2 NBs and the TiO_2 NBs protruding from TiO_2 NPs using a hydrothermal approach, and interconnected these NBs into networks. The networks showed better conductivity than that of the TiO_2 NP film. Using the interconnected TiO_2 NB network as the dye absorber and electron-transport media, the conversion efficiency of DSSC was enhanced.

EXPERIMENTAL SECTION

Synthesis

TiO_2 NPs (99.8 %, Aldrich) were converted to TiO_2 NBs through a hydrothermal process in 15 M KOH water solution and in a sealed Teflon-lined autoclave. A typical reaction for the growth would last 12-36 h at 185 °C. The products were washing with 0.1 M nitric acid and deionized water until the pH value of the washings reaching 7.

Characterization

The morphology and the internal structure of the samples were analyzed using an FEI Tecnai F-20 Transmission Electron Microscope (TEM) and a FEI Sirion field emission Scanning Electron Microscope (SEM) equipped with an energy-dispersive X-ray (EDX) spectrometer. Raman measurement was carried out at room temperature on a HORIBA Jobin Yvon LabRAM HR Raman Microscope using a laser with a wavelength of 532 nm as the excitation source. The conductivity of the films, which consisted of the TiO_2 NPs, the interconnected TiO_2 NBs and the interconnected TiO_2 NBs protruding from the TiO_2 NPs correspondingly, was measured using an Agilent 4156C precision semiconductor parameter analyzer and a probe station operated at room temperature.

Solar cell measurement

The photoanodes were soaked in 0.5 mM N719 dye ($RuL_2(NCS)_2$]: 2 TBA, L = 2,2'-bipyridyl-4,4'-dicarboxylic acid TBA = tetra-n-butylammonium, Dyesol) ethanol solution for 24 h. The photoanodes were washed once with ethanol and then used for photovoltaic measurements. The redox electrolyte was composed of 0.1 M 1-Aminopyridinium iodide (97 %, Aldrich), 0.1 M KI and 0.05 M I_2 in acetonitrile. The counter electrodes were Pt-coated fluorine-doped SnO_2 transparent conducting (FTO) glass substrates (Tec 8, sheet resistance 9 Ω/square, Pilkington). As-fabricated cells were measured by a source meter (Model 2400, Keithley) under simulated AM 1.5 G sunlight (150 W Model 96000, Oriel).

RESULTS AND DISCUSSION

Figure 1. (a) and (b) are TEM images of the interconnected TiO$_2$ NB networks mixed with and without the unreacted TiO$_2$ NPs, respectively. The inset of (b) is corresponding high resolution TEM image. (c) and (d) are SEM images corresponding to the samples shown in (a) and (b) after annealing.

Figure 1a shows that some NBs were synthesized and aligned on the TiO$_2$ NPs after 12 h. The growth of TiO$_2$ NB can be described as following, the erosion of TiO$_2$ NPs resulted in the formation of [Ti(OH)$_6$]$^{2-}$ monomers in the alkaline solution. As in a saturated state, the adjacent [Ti(OH)$_6$]$^{2-}$ monomers were rearranged to the TiO$_2$ NB by oxolation or olation.[11] The alignment of these TiO$_2$ NBs on TiO$_2$ NP was attributed to the oxolation or olation between [Ti(OH)$_6$]$^{2-}$ monomers and the Ti–OH bonds on TiO$_2$ NP surface which had monoatomic-height step defects after partial erosion in the alkaline solution.[12] Our study also suggests that the continuous erosion of TiO$_2$ NPs and oxolation or olation of adjacent [Ti(OH)$_6$]$^{2-}$ monomers leads to the elongation and expansion of TiO$_2$ NBs. When the reaction time reached 48 h, all TiO$_2$ NPs were converted to single crystalline TiO$_2$ NBs along (100) (anatase TiO$_2$), as shown in Figure 1b. Energy-dispersive X-ray (EDX) analysis confirms that the aforementioned structures have same chemical composition of Ti and O. The TiO$_2$ NBs or the TiO$_2$ NBs protruding from TiO$_2$ NPs were then dispersed in deionized water. The solution was filled into the cube constructed on top of the FTO substrate with 70 μm-thick planar adhesive tapes. After drying and then annealing at 400 °C for 2 h, the adjacent TiO$_2$ NBs were interconnected by oxolation or olation. Figure 1c and 1d show the TiO$_2$ NB networks mixed with and without the unreacted TiO$_2$ NPs correspondingly.

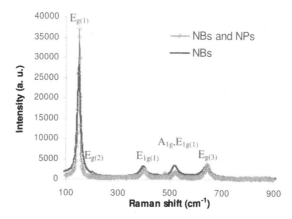

Figure 2. Ambient pressure Raman spectra of the interconnected networks consisting of the TiO$_2$ NBs and of the TiO$_2$ NBs protruding from the TiO$_2$ NPs recorded by a HR800 Raman system with a laser wavelength of 532 nm.

The interconnected networks consisting of the TiO$_2$ NBs or of the TiO$_2$ NBs protruding from TiO$_2$ NPs were further characterized by Raman spectroscopy as shown in Figure 2. All Raman peaks of both samples were related to the typical six Raman-active modes ($3E_g + 2B_{1g} + A_{1g}$) of octahedral anatase TiO$_2$.[13] It indicates that both of these networks are mainly anatase phase. Comparing with the Raman spectrum of the interconnected networks consisting of the TiO$_2$ NBs protruding from TiO$_2$ NPs, the $E_{g(1)}$ peak of the TiO$_2$ NB interconnected networks is lower, and its Raman peaks broadened. In terms of that the size effect on these Raman shifts has been interpreted by the difference between the well confined phonon in a perfect "infinite" crystal and the anharmonic coupling phonons in NP.[14] This result is consistent to that the even size of the TiO$_2$ nanocrystals reduced as all TiO$_2$ NPs were converted to thin TiO$_2$ NBs.

The conductivity of the films consisted of the TiO$_2$ NPs, the interconnected TiO$_2$ NBs or the interconnected TiO$_2$ NBs protruding from TiO$_2$ NPs, were measured at room temperature. Their measured electrical resistivity were 2.6 ×10^7, 6.5×10^3 and 1.5×10^4 Ω•cm, respectively. It indicates that the replacement of the TiO$_2$ NPs with the interconnected TiO$_2$ NBs enhanced the conductivity of the TiO$_2$ film. The electron diffusion through the TiO$_2$ NP film is slow due to the boundaries of the TiO$_2$ NPs and the localized surface electron traps of Ti (III)–OH related to oxygen vacancies, which are located at the step edges rather than embedded in a plateau (such as the smooth surface of the TiO$_2$ NB). For the film of interconnected TiO$_2$ NBs, the boundaries were significantly eliminated and the face to face interconnections between the TiO$_2$ NBs were firm after the oxolation or olation during annealing process. Furthermore, the surface electron traps were reduced by forming high quality TiO$_2$ NB. Therefore the conductivity of the interconnected TiO$_2$ NB film is higher than that of the TiO$_2$ NP film. We need mention here another effect of hopping protons through a percolation cluster formed by bonded water molecules on the NB surfaces,[15] which will be further studied.

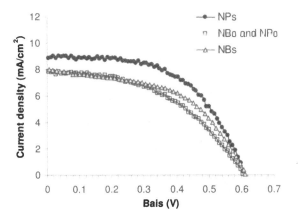

Figure 3. *J-V* plots of as-fabricated DSSCs under 1 sun. TiO$_2$ NP cell: V_{oc} 0.61 V, J_{sc} 7.99 mA/cm^2, FF 0.49, η_{PCE} 2.49 %; TiO$_2$ NB/NP cell: V_{oc} 0.60 V, J_{sc} 7.84 mA/cm^2, FF 0.46, η_{PCE} 2.20 % and TiO$_2$ NB cell: V_{oc} 0.60 V, J_{sc} 8.86 mA/cm^2, FF 0.56, η_{PCE} 3.01 %.

Three 5 μm thick films consisted of the TiO$_2$ NPs, the interconnected TiO$_2$ NBs and the interconnected TiO$_2$ NBs protruding from TiO$_2$ NPs were employed to fabricate DSSCs. Figure 3 shows the *J-V* plots of as-fabricated cells under simulated sunlight with an intensity of AM 1.5 G. The open-circuit voltages (V_{oc}) of three cells were lower than the value reported.[4] It maybe caused by our fabrication process. The V_{oc} of the three cells were similar, while the short-circuit current density (J_{sc}) of the cell consisting of the interconnected TiO$_2$ NB network was higher than that of the TiO$_2$ NP based cell or that of the cell consisting of the interconnected TiO$_2$ NBs protruding from TiO$_2$ NPs. It suggests that the series resistance of the TiO$_2$ porous layer decreased by replacement of the TiO$_2$ NP film with the interconnected TiO$_2$ NB networks.

CONCLUSIONS

We fabricated interconnected networks using the TiO$_2$ NBs synthesized via an alkaline hydrothermal process. This nanostructure was utilized to replace the TiO$_2$ NP film in order to improve the transport of injected electrons. The enhancement of J_{sc} suggests this nanostructure can serve as effective dye absorber and excellent electron-transport media. This study demonstrates a facile method to enhance the DSSC's performance using novel nanostructures.

ACKNOWLEDGMENTS

This work was supported by the National Science Foundation under grants ECCS-0348277 and ECCS-0520891, and a grant from the ONAMI/DOD's nanoelectronic program.

REFERENCES

1. B. O'Regan, M. Grätzel, *Nature* **353**, 737 (1991).
2. N. Robertson, *Angew. Chem. Int. Ed.* **45**, 2338 (2006).
3. D. B. Kuang, P. Wang, S. M. Zakeeruddin, M. Grätzel, *J. Am. Chem. Soc.* **128**, 7732 (2006).
4. T. Horiuchi, H. Miura, K. Sumioka, S. Uchida, *J. Am. Chem. Soc.* **126**, 12218 (2004).
5. M. Dürr, A. Schmid, M. Obermaier, S. Rosselli, A. Yasuda, G. Nelles, *Nat. Mater.* **4**, 607 (2005).
6. L. Kavan, M. Grätzel, S. E. Gilbert, C. Klemenz, H. J. Schell, *J. Am. Chem. Soc.* **118**, 6716 (1996).
7. N. Kopidakis, K. D. Benkstein, J. van de Lagemaat, A. J. Frank, *J. Phys. Chem. B* **107**, 11307 (2003).
8. J. Jiu, S. Isoda, F. Wang, M. Adachi, *J. Phys. Chem. B* **110**, 2087 (2006).
9. B. Tan, Y. Wu, *J. Phys. Chem. B* **110**, 15932 (2006).
10. M. Law, L. E. Greene, J. C. Johnson, R. Saykally, P. Yang, *Nature Mater.* **4**, 455 (2005).
11. Y. Wang, G. Du, H. Liu, D. Liu, S. Qin, N. Wang, C. Hu, X. Tao, J. Jiao, J. Wang, Z. L. Wang, *Adv. Funct. Mater.* **18**, 1131 (2008).
12. X. Gong, A. Selloni, M. Batzill, *Nature Mater.* **5**, 665 (2006).
13. I. R. Beattie, T. R. Gilson, *Proc. R. Soc. London, Ser. A* **307**, 407 (1968).
14. T. Ohsaka, F. Izumi, Y. J. Fujiki, *Raman Spectrosc.* **7**, 321 (1978).
15. G. Garcia-Belmonte, V. Kytin, T. Dittrich, J. Bisquert, *J. Appl. Phys.* **94**, 5261 (2003).

Poster Session:
Compound Semiconductors for Lighting, Power and Sensing

Mater. Res. Soc. Symp. Proc. Vol. 1167 © 2009 Materials Research Society   1167-O04-01

Characterization and Fabrication of InGaN-Based Blue LED With Underlying AlGaN/GaN SLS Cladding Layer Grown on Si(111) Substrate

Bin Abu Bakar Ahmad Shuhaimi[1,2,*], Pum Chian Khai[1,2], Takaaki Suzue[1,2], Yukiyasu Nomura[1,2] and Takashi Egawa[1,2]
[1]Research Center for Nano-Device and System, Nagoya Institute of Technology, Nagoya, Japan
[2]Graduate School of Engineering Physics, Electronics and Mechanics, Nagoya Institute of Technology, Nagoya, Japan

ABSTRACT

This paper reports improved optical characteristics of InGaN-based light-emitting-diode (LED) grown on Si(111) substrate by the insertion of an $Al_{0.06}Ga_{0.94}N$/GaN strained-layer-superlattices (SLS) cladding layer after AlN/GaN multilayer (ML) growth, under the multi-quantum-well (MQW) active layer. The insertion of underlying $Al_{0.06}Ga_{0.94}N$/GaN SLS cladding layer has shown to improve epitaxial layer quality in x-ray diffraction (XRD) analysis, reduce wavelength peak fluctuations in photoluminescence (PL) surface mapping, and improve optical and electrical characteristics of the LED sample. A 34% increase of light intensity at 50 mA current injection and a narrower wavelength peak have been achieved by the insertion of $Al_{0.06}Ga_{0.94}N$/GaN SLS cladding layer. LED with underlying $Al_{0.06}Ga_{0.94}N$/GaN also shows superior current-voltage (I-V) characteristics with operation voltage of 3.2 V at 20 mA and series resistance of 16 Ω.

INTRODUCTION

Si(111) is a very promising substrate for the growth of low-cost GaN-based energy-saving light-emitting devices with wavelength ranging from ultra-violet to infrared illumination. However, large thermal mismatch of 116% and lattice mismatch of 17% between GaN epitaxial layer and the Si(111) substrate contributes to low-quality film with threading dislocation density (TDD) of approximately larger than $\sim 10^{10}$ (cm^{-2}) in conventional metal organic chemical vapor deposition (MOCVD) growth method [1]. The smaller band-gap of Si substrate compared to the band-gap of InGaN-based active layer in visible light-emitting devices structure attributes to the absorption of emitted light by the substrate, causing poor light emission from the device. Ishikawa *et al.* has reported an improved light output power for light-emitting diode (LED) grown in Si(111) substrate using an $Al_{0.3}Ga_{0.7}N$/AlN distributed Bragg reflector (DBR) to reduce the optical loss [2]. However, the pair number is limited to 5 to obtain a crack-free film. Recently, we have reported reduction of TDD and enhancement of internal quantum efficiency in InGaN-based multi-quantum-well with the insertion of $Al_{0.06}Ga_{0.94}N$/GaN SLS cladding layer under the active layer [3]. The $Al_{0.06}Ga_{0.94}N$/GaN SLS cladding layer can also be used to reduce optical loss to the substrate in LED operation. In this paper, we report material characterization and fabrication of InGaN-based blue LED with underlying $Al_{0.06}Ga_{0.94}N$/GaN SLS cladding layer

---

*Corresponding author, e-mail: shuhaimi@msn.com

grown on Si(111) substrate. This paper is intended to demonstrate the superior optical and electrical characteristics that can be achieved by the insertion of underlying $Al_{0.06}Ga_{0.94}N$/GaN SLS cladding layer in LED structure grown on Si(111) substrate.

EXPERIMENT

The LED structure in this study was grown on 2 inch Si(111) substrate by horizontal-reactor MOCVD. Trimethylgallium (TMGa), trimethylaluminum (TMAl) and ammonia (NH_3) were used as precursors for Ga, Al and N, respectively. Hydrogen (H_2) was used as carrier gas. Monosilane (SiH_4) diluted in hydrogen was used for n-type dopant. The substrate was thermally cleaned at 1100°C in H_2 flow before growth. Prior to the growth of LED layers, a 20 nm thin high-temperature AlN layer was grown on the substrate as seeding-layer, followed by 40 pairs of AlN/GaN multilayers (MLs) with respective thickness of 5 nm and 20 nm. Subsequently, a 400 nm thick $Al_{0.06}Ga_{0.94}N$/GaN SLS cladding layer with respective thickness of 2.5 nm each was grown on the MLs. The growth was continued with 200 nm thick n^+-GaN contact layer, active layer consisting of 15 pairs of 2 nm thick $In_{0.16}Ga_{0.84}N$ well and 9 nm thick $In_{0.08}Ga_{0.92}N$ barrier layer, a 10 nm thick p-$Al_{0.08}Ga_{0.92}N$ electron-block layer, and a 50 nm thick p-GaN contact layer for the LED structure. A conventional sample grown with the same parameter but without the $Al_{0.06}Ga_{0.94}N$/GaN SLS cladding layer was also prepared for comparison. The layer structure of the LED is shown in figure 1.

Figure 1. LED structure grown on Si (111) substrate with underlying $Al_{0.06}Ga_{0.94}N$/GaN SLS cladding layer.

Crystal properties of the samples were evaluated by Philips X'Pert Pro x-ray diffraction (XRD) system using hybrid double-axis diffractometer. Photoluminescence (PL) mapping was performed at room-temperature to evaluate the emission characteristics of the samples. The samples were fabricated into LED using standard device processing method reported elsewhere with Ni/Au semi-transparent electrode, Ni/Au p-type contact electrode, Ti/Al/Ni/Au top n-type contact electrode on the n^+-GaN contact layer and AuSb/Au backside n-type contact electrode on the n^+-type Si substrate [4]. Optical characteristics were measured at room temperature using an integrated sphere and electrical characteristics were measured using a standard semiconductor parameter analyzer.

DISCUSSION

All samples in this study are crack-free, showing specular surface and free from melt-back etching. Crystal quality is evaluated by x-ray rocking curve (XRC) ω scan of GaN, with symmetric (0004) scan associated with screw and mixed dislocations, while asymmetric ($20\bar{2}4$) scan is associated with edge and mixed dislocations. The XRC properties of the samples in this study are shown in table I. The full-width-at-half-maximum (FWHM) results in our samples are relatively larger than results reported elsewhere by other groups due to the thin existence of pure GaN film in our structure. However, the comparison between samples in this study shows obvious improvement of crystal quality when SLS cladding layer is inserted in the structure. The (0004) scan result indicates that a lower screw dislocations can be obtained by the insertion of $Al_{0.06}Ga_{0.94}N$/GaN SLS underlying cladding layer. The result also indicates a small improvement of edge dislocations in the structure.

Table I. XRC properties in LED structures with and without underlying SLS.

LED Sample		FWHM (arcsec)	Intensity (a. u.)
With $Al_{0.06}Ga_{0.94}N$/GaN SLS	(0004)	884.8	28400.3
	($20\bar{2}4$)	1826.2	563.8
Without $Al_{0.06}Ga_{0.94}N$/GaN SLS	(0004)	1178.4	7498.6
	($20\bar{2}4$)	1889.7	332.7

Photoluminescence (PL) peak wavelength mapping for the LED structure is shown in figure 2, with (a) for sample with underlying $Al_{0.06}Ga_{0.94}N$/GaN SLS cladding layer, and (b) for conventional sample without the $Al_{0.06}Ga_{0.94}N$/GaN SLS cladding layer. Average MQW peak wavelength in (a) is 437.8 nm, and (b) is 471.9 nm. The $Al_{0.06}Ga_{0.94}N$/GaN SLS cladding layer reduces MQW wavelength peak fluctuations, with standard deviation value of 5.3% in (a) compared to 7.6% in (b). This result indicates that the $Al_{0.06}Ga_{0.94}N$/GaN SLS cladding layer is also effective to improve MQW wavelength peak fluctuations.

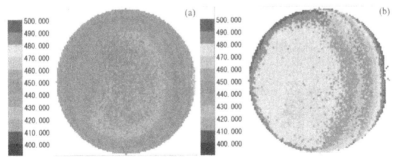

Figure 2. Photoluminescence (PL) wavelength peak mapping for LED structure of (a) with underlying $Al_{0.06}Ga_{0.94}N$/GaN SLS cladding layer and (b) conventional sample. Standard deviation in (a) is 5.3%, and (b) is 7.6%.

Figure 3 shows light intensity vs. current (*I-L*) characteristics for the samples. The results clearly show that LED structure with underlying $Al_{0.06}Ga_{0.94}N$/GaN SLS cladding layer yields a higher intensity and a higher saturation current compared to the conventional structure. At 50 mA current injection, the light intensity from LED with underlying $Al_{0.06}Ga_{0.94}N$ SLS cladding layer is 34% higher than that of from conventional LED structure. In both structures, current injection from the top p-GaN contact layer to the backside n[+]-type Si substrate shows a higher saturation current due to a better current spreading by the substrate, which also spreads heat uniformly in the LED chip.

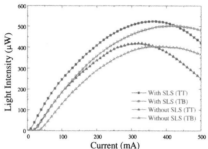

Figure 3. Light intensity vs. current (*I-L*) characteristics for LED with underlying $Al_{0.06}Ga_{0.94}N$/GaN SLS cladding layer and conventional sample. (TT indicates current injection from top p-GaN to top n[+]-GaN layer, while TB indicates current injection from top p-GaN to backside n[+]-type Si substrate).

Electroluminescence (EL) characteristics of the samples are shown in figure 4. The LED structure with underlying $Al_{0.06}Ga_{0.94}N$/GaN SLS cladding layer in (a) shows a narrower spectrum full-width-at-half maximum (FWHM) peak of 27 nm compared to that of 35 nm in conventional sample shown in (b). The narrower FWHM in sample with $Al_{0.06}Ga_{0.94}N$/GaN SLS cladding layer agrees well with the lower standard deviation result in PL wavelength peak mapping for the structure, suggesting lower In composition fluctuation can be achieved in the InGaN-based MQW by the underlying SLS. The narrower FWHM is also attributed by superior

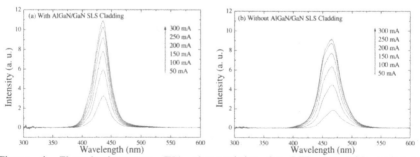

Figure 4. Electroluminescence (EL) characteristics for LED (a) with underlying $Al_{0.06}Ga_{0.94}N$/GaN SLS cladding and (b) for conventional sample.

crystal quality in the sample with underlying $Al_{0.06}Ga_{0.94}N/GaN$ SLS cladding layer compared to that of conventional LED structure, as discussed earlier.

The forward-bias current-voltage (I-V) characteristics are shown in figure 5. A tremendous improvement is seen in top-to-top (TT) I-V characteristics for sample with underlying $Al_{0.06}Ga_{0.94}N/GaN$ SLS cladding layer, with operating voltage of 3.2 V at 20 mA current and series resistance of 16 Ω. This significant improvement is due to lower dislocation density in the sample. The top-to-bottom (TB) I-V characteristics of the same sample shows higher operating voltage of 3.6V at 20 mA and series resistance of 29 Ω, which is influenced by the highly resistive AlN/GaN MLs in the buffer layer. The conventional structure shows almost similar characteristics as InGaN-based LED grown on AlN/sapphire template reported by our group elsewhere [5].

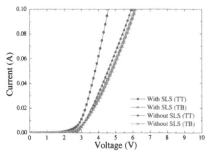

Figure 5. Current vs. voltage (I-V) characteristics for LED with underlying $Al_{0.06}Ga_{0.94}N/GaN$ SLS cladding layer and conventional sample. (TT indicates current injection from top p-GaN to top n^+-GaN layer, while TB indicates current injection from top p-GaN to backside n^+-type Si substrate).

CONCLUSIONS

This study shows that the insertion of $Al_{0.06}Ga_{0.94}N/GaN$ SLS cladding layer with a low composition of Al under the active layer is effective to improve crystal quality, reduce MQW wavelength peak fluctuations, and improve overall optical and electrical characteristics of the LED. Light reflectance from the underlying cladding layer has also improved the LED optical characteristics with a higher light intensity and a narrower FWHM spectrum.

REFERENCES
1. T. Egawa, B. Zhang, N. Nishikawa, H. Ishikawa, T. Jimbo and M. Umeno, *J. Appl. Phys.* **91**, 528 (2002).
2. H. Ishikawa, K. Asano, B. Zhang, T. Egawa and T. Jimbo, *Phys. Stat. Sol. (a)* **201**, 2653 (2004).
3. B. A. B. Ahmad Shuhaimi, T. Suzue, Y. Nomura, Y. Maki and T. Egawa, *MRS Spring Meeting 2009* **O2.5**, San Francisco (2009).
4. B. J. Zhang, T. Egawa, H. Ishikawa, N. Nishikawa, T. Jimbo and M. Umeno, *Phys. Stat. Sol. (a)* **188**, 151 (2001).
5. B. Zhang, T. Egawa, Y. Liu, H. Ishikawa and T. Jimbo, *Phys. Stat. Sol. (c)* **0**, 2244 (2003).

Mater. Res. Soc. Symp. Proc. Vol. 1167 © 2009 Materials Research Society 1167-O04-07

Growth by Molecular Beam Epitaxy of GaNAs alloys with high As content for potential photoanode applications in hydrogen production

Sergey V. Novikov[1], Chris R. Staddon[1], Andrey V. Akimov[1], Richard P. Campion[1], Norzaini Zainal[1], Anthony J. Kent[1], C. Thomas Foxon[1], Chien H. Chen[2], Kin M. Yu[3] and Wladek Walukiewicz[3]

[1]School of Physics and Astronomy, University of Nottingham, Nottingham NG7 2RD, UK
[2]Photovoltaics Technology Center, Industrial Technology Research Institution, Taiwan 310, Republic of China
[3]Materials Sciences Division, Lawrence Berkeley National Laboratory, 1 Cyclotron Road, Berkeley, CA 94720-8197, U.S.A.

ABSTRACT

We have studied the low-temperature growth of GaNAs layers on sapphire substrates by plasma-assisted molecular beam epitaxy. We have succeeded in achieving $GaN_{1-x}As_x$ alloys over a large composition range by growing the films at temperature much below the normal GaN growth temperatures with increasing the As_2 flux as well as Ga:N flux ratio. We found that the alloys with high As content $x>0.1$ are amorphous. Optical absorption measurements reveal a continuous gradual decrease of band gap from ~3.4 eV to ~1.4 eV with increasing As content. The energy gap reaches its minimum of ~1.4 eV at the x~0.6-0.7. For amorphous GaAsN alloys with $x<0.3$ the composition dependence of the band gap follows the prediction of the band anticrossing model developed for dilute alloys. This suggests that the amorphous $GaN_{1-x}As_x$ alloys have short-range ordering that resembles random crystalline $GaN_{1-x}As_x$ alloys. Such amorphous $GaN_{1-x}As_x$ alloys with tunable electronic structure may be useful as photoanodes in photo-electrochemical cells for hydrogen production.

INTRODUCTION

Photoelectrochemical (PEC) cells, illuminated by sunlight, have the ability to split water into hydrogen and oxygen [1]. Such cells use photoactive electrodes immersed in an aqueous electrolyte or water. The choice of material for the photoanode (photocathode) is crucial for efficient hydrogen production using the PEC method. The band gap of semiconductor materials used for photoanodes must be at least 1.8-2.0 eV [1] but small enough to absorb most sunlight. In addition to choosing the correct band gap, the conduction and valence band edges must "straddle" the H^+/H_2 and O_2/H_2O redox potentials so that spontaneous water splitting can occur. The second requirement is for the photoanode material to be corrosion-resistant in water solutions for prolong operation. Gallium nitride (GaN) is a good candidate for this application since it has a band gap ~3.4 eV, high mechanical hardness and high chemical stability [2]. Moreover, the band gap of GaN can be adjusted and decreased due to strong negative bowing in the GaN-based solid solutions with group V elements [3].

Recently, it has been demonstrated that the electronic structure of the conduction or valence bands of alloys with anions of very different electronegativity can be well described by the band anticrossing (BAC) model [4]. Due to the substantial difference in the ionization energies of the As and N anions, the p-states of the substitutional As atoms are resonant near the valence band edge of

GaN, allowing them to interact strongly with the extended p-states of the host. The interaction splits the valence band into a series of E_+ and E_- sub-bands with dispersion relations that are dependent on the As concentration [3,4]. As a result, this effectively reduces the band gap and pushes the valence band edge up toward the O_2/H_2O potential, while leaving the conduction band above the H^+/H_2 potential. Therefore, we suggest that that the $GaN_{1-x}As_x$ material system is one of the most promising materials for the photoanodes. On the other hand, in the As-rich alloys, the hybridization of the localized N s-states with the GaAs extended s-states a strong decrease in the conduction band edge. These two effects cause the large bowing across the entire composition range of the Ga-N-As alloy system resulting in energy gap ranging from ~0.7 to 3.4 eV [3].

In the last decade there has been considerable theoretical and experimental interest in As-doped GaN [5]. The interest in As doped GaN has been motivated by three effects: the abrupt decrease in the band gap for GaNAs solid solutions, blue emission at room temperature from arsenic doped GaN and arsenic-stimulated growth of the metastable cubic (zinc-blende) phase of GaN. Previously, we have reported in great detail our investigations on the growth and properties of GaNAs layers prepared by molecular beam epitaxy (MBE) [5]. However, a large miscibility gap was theoretically predicted and experimentally confirmed for the Ga-N-As system. $GaN_{1-x}As_x$ alloys at the N-rich end of the phase diagram have been grown by metal-organic vapour phase epitaxy (MOVPE) and by MBE [see references in [5]]. For both techniques, it is difficult to obtain a high concentration of As in the alloy before phase separation occurs. The highest concentrations reported in MOVPE layers is x~0.067 [3,6] and in MBE layers are x~0.0026 [7] grown at 750°C and x~0.01 [8] at 500°C respectively. Since it is known that As solubility limit in $GaN_{1-x}As_x$ alloys increases with decreasing growth temperature [8,9] we investigate the growth of $GaN_{1-x}As_x$ alloys by MBE at low temperature.

EXPERIMENT

All GaNAs samples were grown on 2" sapphire (0001) substrates by plasma-assisted MBE (PA-MBE) in a MOD-GENII system. The system has a HD-25 Oxford Applied Research RF activated plasma source to provide active nitrogen and elemental Ga is used as the group III-source. In all experiments we have used arsenic in the form of As_2 produced by the Veeco arsenic valved cracker. The MBE system is equipped with reflection high energy electron diffraction (RHEED) facility for surface reconstruction analysis. For the growth of all GaNAs samples, we have used the same active N flux (total N beam equivalent pressure (BEP) ~1.5 10^{-5} Torr) and the same deposition time (2 hr) for the majority of the films. In this paper we present the temperature measured by the substrate heater thermocouple as a reference temperature knowing that this is not the real temperature of the substrate surface.

We have used a wide range of *in-situ* and *ex-situ* characterisation techniques to study the surface morphology, film composition, structural, electrical and optical properties of $GaN_{1-x}As_x$ layers. The morphology of the samples was studied *in-situ* using RHEED and *ex-situ* using atomic force microscopy (AFM). The structure and orientation of the GaNAs layers was studied using X-ray diffraction (XRD) using the Philips X'Pert MRD diffractometer. The As content in the GaNAs films was determined by combined Rutherford backscattering spectrometry (RBS) and particle-induced x-ray emission (PIXE) measurements using a 2 MeV $^4He^+$ beam. Note that the As composition measured by RBS/PIXE is the overall As content in the films but not necessarily As substituting the N sublattice. The optical properties of the $GaN_{1-x}As_x$ layers were studied by reflection and absorption measurements in the wavelength range of 200-1100 nm.

RESULTS AND DISCUSSION

We first investigate the growth of GaNAs alloys at relatively high temperatures of ~500-600°C with a high As overpressure. As_2 flux with a beam equivalent pressure (BEP) 7 10^{-6} Torr was used. For the GaNAs layers grown under a slightly Ga rich condition we observe phase separation into GaN:As and GaAs:N. The XRD pattern of these films show both the (0002) diffraction peaks of GaN doped with As at 2θ~35° and the (111) peak of GaAs doped with N [7] at 2θ~27°.

Results are different for $GaN_{1-x}As_x$ films grown at the same temperature ~500-600°C, but under N-rich conditions. The XRD pattern shows only the (0002) diffraction peak of a GaNAs alloy, suggesting that a single phase $GaN_{1-x}As_x$ alloy is formed. The (0002) peak of the $GaN_{1-x}As_x$ alloys is clearly shifted to lower angle in comparison with GaN, suggesting that the lattice parameter of the alloy is larger than that of GaN due to the substitution of a small fraction of As in the N sublattice. Using Vegaard's law and the calculated lattice parameter of hexagonal GaAs [10], the GaAs mole fraction can be estimated to be x~0.015. This is in good agreement with the As composition of x~0.02 estimated by RBS/PIXE measurements.

It is well known that the solubility of impurities increases with decreasing growth temperature and at extremely low growth temperature composition far exceeding the equilibrium solubility limit can be attained. In order to increase the As content in $GaN_{1-x}As_x$ alloys we carried out growth studies under N-rich conditions for a wide temperature range from ~600°C to ~100°C. RHEED patterns show that with decreasing growth temperature, the GaNAs layers become amorphous. This was later confirmed by X-ray studies. AFM studies show that the surface of low temperature grown amorphous GaNAs samples are extremely smooth with an RMS roughness ~0.2 nm. RBS/PIXE studies estimate that the As concentration inside the amorphous GaNAs layers gradually increases from a few percent at ~500°C to x~0.70 at the lowest growth temperature of ~100°C.

Optical absorption and reflection measurements show a progressive shift of the optical band gap to lower energy with decreasing growth temperature. The square of the absorption coefficient α^2 versus photon energy for $GaN_{1-x}As_x$ films grown at different temperatures are shown in figure 1. The direct band gap of the alloy can be estimated by extrapolating the linear part of the absorption edge down to the energy axis. The overall As content measured by RBS/PIXE are also given in the figure. Figure 1 shows the samples grown in the temperature range of 100-600°C with an As_2 flux BEP ~7 10^{-6} Torr. Despite the fact that GaNAs layers are amorphous the energy gap decreases monotonically as the growth temperature decreases.

We have also grown a set of GaNAs samples at low growth temperature ~300°C with As_2 flux ranging from 0 to 7 10^{-6} Torr. GaNAs layers grown without an As flux or with a very low As flux are transparent and crystalline. However, at higher As flux, GaNAs layers become amorphous and only semi-transparent. For low As doped GaNAs layers, we see a gradual shift of (0002) GaN reflection towards lower angle in X-ray studies, which suggests that we are forming $GaN_{1-x}As_x$ alloys. A gradual increase of the As concentration in the $GaN_{1-x}As_x$ layers up to x~0.50 with increasing As_2 flux was observed. Optical absorption results on this set of samples are shown in figure 2. Results in figure 1 and figure 2 indicate that lowering the growth temperature or increasing As$_2$ flux increases As content in the layers and shifts the optical band gap to lower energy.

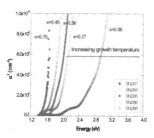

Figure 1. The energy dependence the square of the absorption coefficient α^2 for GaN$_{1-x}$As$_x$ layers, grown at different temperatures from ~100°C to ~600°C.

In PA-MBE growth of GaN:As layers, the Ga to N flux ratio plays a very important role. We have grown another set of GaNAs samples at ~300°C with fixed N and As (~7 10^{-6} Torr) fluxes while changing the Ga flux over a wide range from 7 10^{-8} Torr to 4 10^{-7} Torr. Unlike for high temperature growth, no Ga droplet formation on sample surface is observed even under very Ga-rich conditions. The As content in the GaNAs layers increases with increasing Ga to N ratio from x~0.3 to x~0.6. Optical band gap shifts to lower energy with increasing Ga:N flux ratio.

At low growth temperatures As content as high as x~0.7 can be obtained in GaNAs layers. The lack of any XRD peak in these high As content films suggests that GaNAs films grown at low temperature are amorphous in structure, although nanocrystalline inclusion in an amorphous matrix cannot be ruled out. We note that no evidence for phase separation in these GaNAs layers for all As concentrations can be detected. This agrees with our observation that no distinct absorption edges at close to GaAs or GaN were measured. Instead we observe a monotonic decrease in band gap as a function of As content in these films.

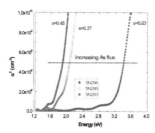

Figure 2. The energy dependence the square of the absorption coefficient α^2 for GaN$_{1-x}$As$_x$ layers grown at different As$_2$ fluxes from 1.2 10^{-8} Torr to 2.7 10^{-6} Torr at ~300°C.

In this study, we are able to increase the incorporation of As in GaN by (i) lowering the growth temperature, (ii) increasing the As$_2$ flux at low growth temperature, and (iii) increasing Ga:N flux ratio. The increase in As content in the film with decreasing growth temperature can be explained by an increase in As surface lifetime on the growing film surface. An increase in As content is also observed with increasing As flux is also due to reduced As re-evaporation at low temperatures. The increase in As content with increasing Ga:N flux ratio must be due to kinetic factors which become more important at low growth temperatures.

Figure 3 plots the optical band gap of the $GaN_{1-x}As_x$ films as a function of the overall As content measured by RBS/PIXE. Note that the samples with phase separation are not included here. Despite the fact that the samples with high As content x>0.1 are amorphous and those with x<0.1 are crystalline, we see a gradual continuous decrease of band gap from ~3.4 eV to ~1.4 eV with increasing As content. The energy gap reaches its minimum of ~1.4 eV at the x~0.6-0.7. This band gap reduction cannot be explained by the amorphous GaN and/or GaAs materials. It has been reported that amorphous GaN and nanocrystalline GaN with crystallites as small as 3 nm exhibit an electronic structure resembling a broadened version of that in crystalline GaN [11,12]. Using the band anticrossing effect in N-rich and As-rich GaNAs alloys, it is possible to interpolate the composition dependence of the band gap of $GaN_{1-x}As_x$ alloys over the entire composition [3] (solid curve in figure 3). Our experimental data from amorphous $GaN_{1-x}As_x$ follow this calculated dependence rather well. Given the fact that the calculated band gap is an interpolation of results from very diluted GaNAs and GaAsN alloys, the agreement is remarkable. However, our measured band gap of amorphous $GaN_{1-x}As_x$ with x >0.3 are significantly larger than that predicted by BAC.

CONCLUSIONS

We have succeeded in growing $GaN_{1-x}As_x$ alloys over a large composition range (0<x<0.7) by plasma-assisted MBE. At relatively high growth temperature of ~600°C, we have grown crystalline $GaN_{1-x}As_x$ with x~0.02. The enhanced incorporation of As was achieved by growing the films at temperature much below the normal GaN growth (>600°C) with increasing the As_2 flux as well as Ga:N flux ratio. In this study we found that the alloys with high As content x>0.1 are amorphous. Optical absorption measurements reveal a continuous gradual decrease of band gap from ~3.4 eV to ~1.4 eV with increasing As content. The energy gap reaches its minimum of ~1.4 eV at the x~0.6-0.7. The composition dependence of the band gap of the amorphous $GaN_{1-x}As_x$ alloys follows the prediction of the band anticrossing model for x<0.3. The results indicate that for this composition range the amorphous $GaN_{1-x}As_x$ alloys have short-range ordering that resembles random crystalline $GaN_{1-x}As_x$ alloys. These $GaN_{1-x}As_x$ alloys over the whole composition range can be used not only for photoanodes applications in PEC cells for hydrogen production, but also have technological potentials for many optical devices operating from the ultraviolet (~0.4 μm) to the infra-red (~2 μm).

ACKNOWLEDGMENTS

This work at the University of Nottingham was undertaken with support from the EPSRC (EP/G007160/1 and EP/D051487/1). The work performed at LBNL was supported by the Director, Office of Science, Office of Basic Energy Sciences, Materials Sciences and Engineering Division, of the U.S. Department of Energy under Contract No. DE-AC02-05CH11231.

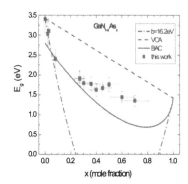

Figure 3. Optical band gap of the GaN$_{1-x}$As$_x$ films as a function of the overall As content measured by RBS/PIXE. Calculated composition dependence of the band gap of GaN$_{1-x}$As$_x$ alloys based on the band anticrossing model (BAC), virtual crystal approximation (VCA) and using a single bowing parameter extracted from dilute alloys (b=16.2 eV) are also shown.

REFERENCES

1. C. C. Sorrell, S. Sugihara and J. Nowotny, Eds., *Materials for energy conversion devices* (Woodhead Publishing Limited, Cambridge, England, 2005) ISBN-10:1 85573 932 1 and references therein.
2. J. H. Edgar, S. Strite, I. Akasaki, H. Amano and C. Wetzel, Eds., *Gallium Nitride and Related Semiconductors* (INSPEC, Stevenage, 1999) ISBN 0 85296 953 8.
3. J. Wu, W. Walukiewicz, K. M. Yu, J. D. Denlinger, W. Shan, J. W. Ager, A. Kimura, H. F. Tang and T. F. Kuech, *Phys Rev B* **70**, 115214 (2004).
4. W. Walukiewicz, K. Alberi, J. Wu, W. Shan, K.M. Yu, and J. W. Ager III, in Physics of Dilute III-V Nitride Semiconductors and Material Systems: Physics and Technology, edited by Ayse Erol (Springer-Verlag Berlin-Heidelberg 2008) Chapter 3.
5. C. T. Foxon, I. Harrison, S. V. Novikov, A. J. Winser, R.P. Campion and T. Li, *J. of Physics: Condensed Matter* **14**, 3383 (2002) and references therein.
6. A. Kimura, C. A. Paulson, H. F.Tang and T. F. Kuech, *Appl. Phys. Lett.* **84**, 1489 (2004).
7. K. Iwata, H. Asahi, K. Asami, R. Kuroiwa and S. Gonda, *Jpn. J. Appl. Phys.* **37**, 1436 (1998).
8. Y. Zhao, F. Deng, S. S. Lau and C. W. Tu, *J. Vac. Sci. Technol. B* **16**, 1297 (1998).
9. S. V. Novikov, T. Li, A. J. Winser, R. P. Campion, C. R. Staddon, C. S. Davis, I. Harrison and C. T. Foxon, *Phys. Stat. Sol. (b)* **228**, 223 (2001).
10. C. Y. Yeh, Z. W. Lu, S. Froyen, and A. Zunger, *Phys. Rev. B* **46** 10086 (1992).
11. P. Stumm and D. A. Drabold, *Phys. Rev Lett.* **79**, 677 (1997).
12. B. J. Ruck, A. Koo, U. D. Lanke, F. Budde, H. J. Trodahl, G. V. M. Williams, A. Bittar, J. B. Metson, E. Nodwell, T. Tiedje, A. Zimina and S. Eisebitt, *J. Appl. Phys.* **96**, 3571 (2004).

Compound Semiconductors for Energy

Mater. Res. Soc. Symp. Proc. Vol. 1167 © 2009 Materials Research Society   1167-O05-03

1.2 kV AlGaN/GaN Schottky Barrier Diode Employing As+ Ion Implantation on SiO₂ Passivation Layer

Jiyong Lim, Young-Hwan Choi, Young-shil Kim, Min-Ki Kim and Min-Koo Han
School of Electrical Engineering and Computer Science, Seoul National University,
301-1115 Seoul National University, Shillim-9 dong Gwanak-gu, Seoul, Korea, 151-744

ABSTRACT

We proposed and fabricated 1.2 kV AlGaN/GaN Schottky barrier Diode (SBD) employing As+ ion implantation on SiO₂ passivation layer. As+ ions which had been implanted changed depletion region curvature under negative bias condition, so that the breakdown voltage increased and the leakage current decreased. The breakdown voltage of the proposed device was 1204 V while that of the conventional device was 604 V. The leakage current of the proposed device was 21.2 nA/mm when the cathode bias was -100V while that of the conventional device was 80.3 uA/mm at the same condition. Positive charge of implanted As+ ions induced electrons at the 2 dimensional electron gas (2DEG) region, so that the channel density increased slightly. Thus, forward current increased.

INTRODUCTION

AlGaN/GaN Schottky barrier diode (SBD) and high electron mobility transistors (HEMTs) have attracted considerable attention because of a high breakdown field in the wide band-gap semiconductor [1, 2]. A breakdown voltage is very important in high power applications. Considerable amount of works have been reported to achieve a high breakdown voltage by employing additional processes such as SiO₂ passivation, floating metal rings and Ni/Au Oxidation [3-5]. The breakdown of AlGaN/GaN SBDs is related with the potential distribution between the anode and the cathode electrodes. Passivation layer suppresses electron trapping at surface states. Passivation layer changes the potential distribution between the anode and the cathode electrodes and increases the breakdown voltage [6].

Without a passivation, electrons are trapped from the anode to the surface states when the device is reverse-biased. Then the virtual anode is formed at the anode-cathode access region, so that the high-field region is formed at the cathode edge [6, 7]. In the passivated AlGaN/GaN SBDs, high field at the cathode edge is relaxed. A potential drop occurs at the anode edge, and electric field concentration is terminated due to the suppression of the virtual anode [6].

The purpose of our work is to report an increase of breakdown voltage in AlGaN/GaN SBDs by employing As+ ion implantation on SiO₂ passivation. Treatment of passivation layer has been reported scarcely. We have fabricated AlGaN/GaN SBDs with passivation layer and implanted As+ ion on the surface passivation layer of devices. After 1×10^{14} /cm² dose and 80 keV energy implantation, depletion region curvature was improved so that the maximum electric field strength was decreased so that the breakdown voltage of AlGaN/GaN HEMTs increased and the leakage current of AlGaN/GaN HEMTs decreased. The breakdown voltage of the proposed device was 1204 V while that of the conventional device was 604 V. The leakage current of the proposed device was 21.2 nA/mm when the cathode bias was 100V while that of the conventional device was 80.3 uA/mm at the same condition.

Experimental Results and Discussion

20 µm

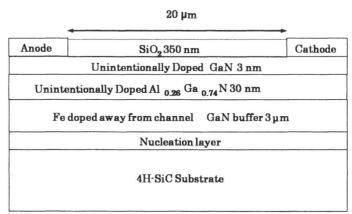

Anode	SiO₂ 350 nm	Cathode
Unintentionally Doped GaN 3 nm		
Unintentionally Doped Al₀.₂₆Ga₀.₇₄N 30 nm		
Fe doped away from channel GaN buffer 3 µm		
Nucleation layer		
4H-SiC Substrate		

Fig. 1. Cross-sectional view of the proposed AlGaN/GaN Schottky barrier diode

The cross-sectional view of the proposed AlGaN/GaN Schottky barrier diode is shown in Fig. 1. We fabricated conventional AlGaN/GaN Schottky barrier diode and passivated the device with SiO₂ layer of 350 nm thick. AlGaN/GaN heterostructure was grown on semi-insulating 4H-SiC substrate by metal organic chemical vapor deposition (MOCVD). The 270 nm mesa structure for isolation was formed by inductively coupled plasma etching. Ohmic metal the cathode electrodes, Ti/Al/Ta/Au (20/80/20/100 nm) were formed simultaneously with an e-gun evaporator and defined by a liftoff technique and annealed at 850 °C for 30 s under N₂ ambient. Schottky metal of the anode electrode Ni/Au/Ni (50/300/50 nm) was formed simultaneously with an e-gun evaporator and defined by a lift-off technique. The SiO₂ passivation layer (350 nm) was deposited with inductively coupled plasma chemical vapor deposition (ICP CVD). Finally As+ ions were implanted on the SiO₂ passivation layer.

We calculated the mean penetration depth of various ions. As+ is determined as the most proper ion. Too small ions are not proper because their mean penetration depth is too large to prevent the surface from being damaged. Too large ions are not proper either, because their mean penetration depth is too small to have an effect to the channel. The acceleration energy is was calculated by numerical simulation. The dose was selected to make proper number of traps. Too many traps may make switching speed lower while too small traps cannot be able to make an effect on 2DEG. As+ ion implantation was done with Varian E220 ion implanter.

We measured the surface potential of the test samples with electric force microscopy (EFM) in order to verify that implanted As+ ions remained as positively charged ions in SiO₂ layer after ion implantation. The cross-sectional view of the EFM testing structure and the results of the EFM measurements are shown in Fig. 2.

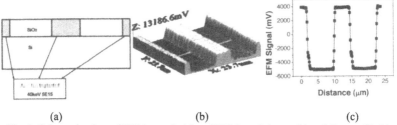

(a)   (b)   (c)

Fig. 2: Schematic view of EFM sample (a) and EFM result image(b) and line profile(c)

After ion implantation, 2 dimensional electron gas (2DEG) concentration was increased slightly from 8.28×10^{12} /cm^2 to 8.38×10^{12} /cm^2 so that the forward current was also increased slightly. Measured forward current is shown in Fig. 3.

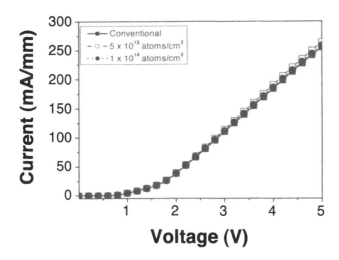

Fig. 3. Measured forward I-V characteristics of the proposed AlGaN/GaN SBDs

Fig. 4 shows the breakdown voltages of the SBDs before and after As+ ion implantation. After 1×10^{14} /cm^2 dose and 80 keV energy implantation, the breakdown voltage increased considerably from 604 V to 1204 V due to the edge termination by implanted As+ ions.

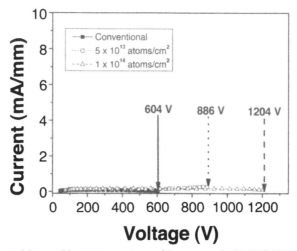

Fig. 4. Measured breakdown voltage of the proposed AlGaN/GaN SBDs

Fig. 5 shows the reverse leakage current of the proposed device. The reverse leakage current decreased from 80.3 uA/mm to 21.2 nA/mm due to the relaxation of electric field concentration by As+ ion implantation.

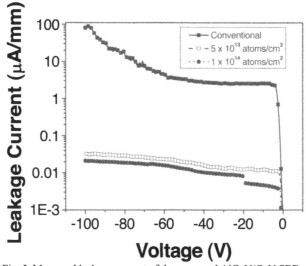

Fig. 5. Measured leakage current of the proposed AlGaN/GaN SBDs

We verified the electric field relaxation through 2D simulation. After As+ ion implantation, the depletion region curvature under the reverse biased condition became moderate so that the maximum electric field strength was decreased as shown in Fig. 6.

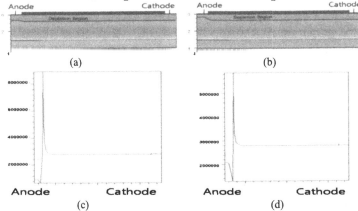

(a)   (b)

(c)   (d)

Fig. 6. TCAD simulation results of conventional SBD (a), (c) and proposed SBD (b),(d)
Vanode-cathode = -100 V

As+ ion implantation method may be a simple and effective edge termination method for improving the breakdown voltage as well as the leakage current of AlGaN/GaN SBDs. Proposed AlGaN/GaN SBDs showed high breakdown voltage of 1204 V and low leakage current of 21.2 nA/mm without any considerable decrease of forward characteristics while that of conventional device was 604 V and 80.3 uA/mm, respectively.

DISCUSSION

The breakdown voltage and the leakage current was improved significantly without sacrificing forward electric characteristic by employing As+ ion implantation on SiO_2 passivation layer. Implanted As+ ions compensated the electrons, which were injected from the anode under negative bias condition, so that the formation of the virtual anode was prevented. Therefore, electric field concentration near the cathode was relaxed so that the electric field of the proposed device was distributed more uniformly than that of the conventional device.

2DEG density of AlGaN/GaN heterostructure has strong relation with surface state. Implanted As+ ions induced electrons at 2DEG region, so that the 2DEG density and the forward current were increased after As+ ion implantation.

CONCLUSIONS

We have fabricated AlGaN/GaN SBDs with passivation layer and implanted As+ ion on the surface passivation layer of devices. After 1×10^{14} /cm^2 dose and 80 keV energy implantation, depletion region curvature was improved so that the maximum electric field strength was decreased so that the breakdown voltage of AlGaN/GaN HEMTs increased and the leakage current of AlGaN/GaN HEMTs decreased. The breakdown voltage of the proposed device was 1204 V while that of the conventional device was 604 V. The leakage current of the proposed device was 21.2 nA/mm when the cathode bias was 100V while that of the conventional device was 80.3 uA/mm at the same condition. As+ ion implantation method may be a simple and effective edge termination method for improving the breakdown voltage as well as the leakage current of AlGaN/GaN SBDs.

ACKNOWLEDGMENTS

This work has been supported by the Proton Engineering Frontier Project of Korea Atomic Energy Research Institute.

REFERENCES

1. S. J. Pearton, J. Appl. Phys., Vol. 86, pp1-78, 1999.
2. Y. –F. Wu, et. al., Appl. Phys. Lett., Vol. 69, pp1438-1440, 1996.
3. S. Arulkumaran, et. al., Appl. Phys. Lett. Vol. 84, pp613-615, 2004
4. M.-W. Ha, et. al., Jpn. J. Appl. Phys., Vol. 44, pp6385-6388, 2005
5. S-C. Lee et. al., Jpn. J. Appl. Phys., Vol. 45, pp 3398-3400, 2006
6. Y. Ohno, et. al., Appl. Phys. Lett., Vol. 84, pp2184-2186, 2004
7. R. Vetury, et. al., IEEE Trans. Electron Devices, Vol. 48, pp560-566, 2001

Mater. Res. Soc. Symp. Proc. Vol. 1167 © 2009 Materials Research Society   1167-O05-06

Diffusion Effect Between Schottky Metals and AlGaN/GaN Heterostructure During High Temperature Annealing Process

Young-Hwan Choi, Jiyong Lim, Young-Shil Kim and Min-Koo Han
School of Electrical Engineering and Computer Science, Seoul National University,
599 Gwanangno, Gwanak-gu, Seoul, 151-744, Korea

ABSTRACT

We have investigated the change of the Schottky contact surface and the interface between Schottky metals and AlGaN/GaN heterostructure after the annealing process for 35 min at 300 °C. The secondary ion mass spectroscopy (SIMS) and the scanning electron microscopy (SEM) show that the Schottky metals and AlGaN/GaN heterostructure interacted actively during the annealing process. The atoms in Schottky contact and AlGaN/GaN heterostructure diffused interactively and the surface roughness of Schottky contact was increased. After the annealing process for fabricated AlGaN/GaN High-Electron-Mobility Transistor (HEMT), the threshold voltage was shifted by +0.2 V and the leakage current was decreased by 40 %.

INTRODUCTION

AlGaN/GaN HEMTs (high-electron-mobility transistors) may be promising devices for microwave and high voltage applications due to wide band gap and low intrinsic carrier density of GaN and two-dimensional electron gas and high saturation velocity of AlGaN/GaN heterostructure [1-2].

AlGaN/GaN HEMTs for high voltage application would require a low leakage current and a high breakdown voltage. The leakage current of AlGaN/GaN devices is due to the poor interface between Schottky metal and AlGaN/GaN heterostructure. When GaN epitaxial structure is grown, dislocations are formed due to the lattice mismatch between GaN and substrate so a considerable amount of defects is formed at the surface of GaN wafer. These defects deteriorate Schottky contact and cause the leakage current of GaN based devices [3]. AlGaN/GaN heterostructure is also grown on the substrate such as SiC, sapphire and Si, so defects induced by the mismatch between GaN and substrate are formed on the surface of wafer.

The high temperature annealing is a simple and effective process for the improvement of leakage current and breakdown voltage of AlGaN/GaN HEMTs [4-6]. It has been reported that the high temperature annealing after Schottky contact metallization reduces the trapping effect on AlGaN surface and/or GaN buffer layer [4] and decreases the density of Schottky metal/AlGaN interface states [5]. The passivation on the AlGaN/GaN HEMTs is also applied for the decrease of the leakage current [7-8]. However, the deposition process employing plasma for the passivation layer requires high temperature condition.

AlGaN/GaN HEMTs are well known for the stable operation in the high temperature condition due to the material characteristics of GaN. But Schottky contact can be changed in high temperature environment and the change of Schottky contact affects the electrical characteristics of AlGaN/GaN devices.

We have investigated the change of the Schottky contact surface and the interface between Schottky metal and AlGaN/GaN heterostructure after the annealing process for 35 min at 300 °C. We measured the interface and the surface of Schottky contact by SIMS and SEM before and after the annealing process. We have fabricated a AlGaN/GaN HEMT and measured the electrical characteristics in order to investigate the effect on the annealing process. The Ni/Au Schottky contact shows the low thermal stability under the 300 $^{\circ}$C condition.

EXPERIMENT

The cross-sectional view of sample A is shown in Fig. 1(a). AlGaN/GaN heterostucture was grown on sapphire substrate. The nucleation layer was grown, followed by the 2 μm-thick Fe doped GaN layer as a buffer. The 20 nm-thick unintentionally doped (UID) $Al_{0.25}Ga_{0.75}N$ layer and the UID GaN capping layer were grown in sequence. Schottky contact, Ni/Au (50/300 nm), was formed by e-gun evaporation. The sample was annealed for 35 min at 300 $^{\circ}$C under N_2 ambient. The bare AlGaN/GaN wafer (sample B) was used for the comparison with sample A. The Sample B was also annealed for 35 min at 300 $^{\circ}$C under N_2 ambient. The cross sectional view of sample B is shown in Fig. 1(b). The nucleation layer was grown on sapphire substrate. The 2.5 μm-thick Fe doped GaN layer as a buffer and the 25 nm-thick unintentionally doped (UID) $Al_{0.25}Ga_{0.75}N$ layer were formed on the nucleation layer. The sample A and sample B were used for SEM and SIMS analysis. The primary ion and energy for the SIMS (magnetic sector type) is O^{2+} and 5.5 kV.

Au	300 nm
Ni	50 nm
Undoped GaN	2 nm
Undoped $Al_{0.25}Ga_{0.75}N$	20 nm
GaN (Fe doped away from channel)	2 μm
Nucleation layer	
Sapphire	

Undoped $Al_{0.25}Ga_{0.75}N$	25 nm
GaN (Fe doped away from channel)	2.5 μm
Nucleation layer	
Sapphire	

(a) Sample A   (b) Sample B

Fig. 1. Cross-sectional views of sample A and B

In order to investigate the effect on the high temperature annealing on AlGaN/GaN based devices, we have fabricated a AlGaN/GaN HEMT. The cross-sectional view of the fabricated AlGaN/GaN HEMT is shown in Fig. 2. The device process was performed on the AlGaN/GaN heterostructure grown on 4H-SiC substrate. A mesa process was performed for device isolation by dry etching. Ti/Al/Ni/Au (20/80/20/100 nm) for source and drain were deposited using an e-gun evaporator and annealed at 870 $^{\circ}$C for 30 sec under N_2 ambient. Then gate, Ni/Au (50/300 nm), was formed by lift-off process. The device was also annealed for 35 min at 300 $^{\circ}$C under N_2 ambient.

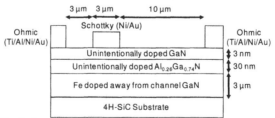

Fig. 2. Cross-sectional views of fabricated AlGaN/GaN HEMT

EXPERIMENTAL RESULTS AND DISCUSSION

The SEM images of the surface of sample A before and after annealing are shown in Fig. 3. The unevenness of sample A before annealing may be attributed to the initial surface of AlGaN/GaN wafer or metal seed formed during the evaporation process. After annealing, the surface of sample A was changed remarkably. It seems that metal was melted and a large number of pits were formed on the metal surface. The roughness of sample A was increased compared with that before the annealing process.

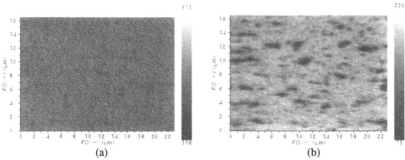

(a)   (b)

Fig. 3. SEM images of the surface of sample A (a) before annealing and (b) after annealing

The depth profiles of sample A before and after annealing were measured by SIMS as shown in Fig. 4. The relative high Ni and Au signals in Al region in Fig. 4 (a) may be due to the unevenness of sample A. After annealing, the depth profile of sample A was changed. Ni and Au were inversed so that Ni was detected from the surface, and they diffused into AlGaN/GaN heterostructure. It should be noticed that Al also diffused and the region of Al after the annealing was wider than that before the annealing, and Ga and Al were detected from the surface of Schottky contact. These results show that the atoms in Schottky contact and AlGaN/GaN heterostructure interact actively during the annealing for 35 min at 300 °C, which results in the change of the interface between Schottky contact and AlGaN/GaN heterostructure.

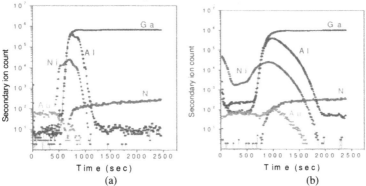

(a)   (b)

Fig. 4. SIMS profiles of sample A (a) before annealing and (b) after annealing

The sample B which does not have Schottky contact was annealed for the comparison with the sample A. The depth profiles of sample B before and after the annealing are almost identical as shown in Fig. 5, which means that Schottky metals such as Ni and Au play an important role to the interaction between AlGaN/GaN heterostructure during high temperature annealing.

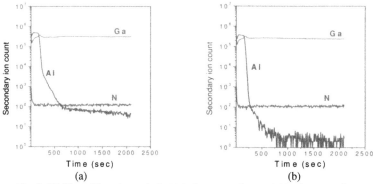

Fig. 5. SIMS profiles of sample B (a) before annealing and (b) after annealing

The electrical characteristics of the fabricated AlGaN/GaN HEMT were measured before and after the annealing. The current between two ohmic contacts did not changed after the annealing process. The transfer characteristics and the forward I-V characteristics are shown in Fig. 6. The drain voltage of 5 V was applied for the measurement of transfer characteristics. After annealing, the transfer curve was shifted to the positive direction and the threshold voltage was increased by 0.2 V. These results may be due to the diffusion of Schottky metals into the AlGaN/GaN heterostructure. The threshold voltage of AlGaN/GaN HEMTs is determined by the distance between Schottky contact and the channel. As the distance is decreased, the threshold voltage moves to the positive direction. We observed the diffusion of Schottky metals into the AlGaN/GaN heterostructure during the annealing process so that the distance between Schottky contact and the channel was decreased and the transfer characteristics of the annealed device were shifted to the positive direction.

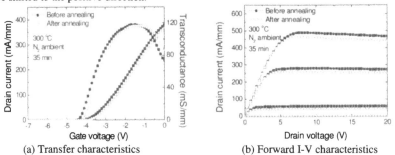

(a) Transfer characteristics   (b) Forward I-V characteristics

Fig. 6. Transfer characteristics and forward I-V characteristics of fabricated AlGaN/GaN HEMT before and after high temperature annealing

The forward I-V characteristics were measured with gate voltage sweep, 1 V to -5 V with -2 V/step. After the annealing process, the on-resistance was identical, but the drain current in the saturation region was decreased by about 5 %. The small decreased drain current is contributed to the positive shift of threshold voltage. As shown in Fig. 6(a), the threshold voltage was increased by 0.2 V so that the drain current measured after the annealing process was smaller compared with that measured before the annealing process at the same bias. Since the applied voltage for the transfer characteristics makes the device operate the saturation region, the drain current in the saturation region was decreased.

The leakage current of the fabricated device before and after the annealing process are shown in Fig. 7. The gate voltage was – 5 V and the drain voltage increased from 0 V to 100 V. The drain current measured at V_{GS}=-5 V and V_{DS}=100 V before the annealing was 3.23 mA/mm, while that after the annealing was 1.92 mA/mm. The leakage current was decreased by 40 %. The improvement of leakage current is due to the reduction the trapping effect on AlGaN surface and/or GaN buffer layer [4] or the decrease of the density of Schottky metal/AlGaN interface states [5].

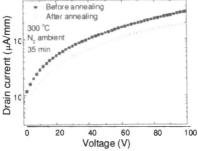

Fig. 7. Leakage current of fabricated AlGaN/GaN HEMT before and after high temperature annealing (V_{GS} = -5 V)

After the annealing for 35 min at 300 °C, the surface roughness of Schottky contact was increased and the atoms in Schottky contact and AlGaN/GaN heterostructure were diffused interactively. Our experimental condition is relatively low at the point of the annealing temperature since the annealing conditions reported were more than 300 °C. But the Schottky contact which consists of Ni and Au shows low stability in that temperature condition. The rough surface of Schottky contact may induce the adhesion problem and decrease the uniformity of fabricated devices. Ni, Au and Al in Schottky contact and AlGaN/GaN heterostructure diffused and were even inverted in some region. The annealing process decreased the leakage current by 40 %. However, it shifted the threshold voltage of fabricated AlGaN/GaN HEMT by +0.2 V, which resulted in the decrease of the saturation current. The shift of threshold voltage is undesirable effect of the high temperature annealing process and it was attributed to the diffusion of Ni or Au.

CONCLUSIONS

We have investigated the change of the Schottky metal surface and the interface between Schottky metals and AlGaN/GaN heterostructure after the annealing process for 35 min at 300 °C. We measured the interface and the surface of Schottky contact by SIMS and SEM before and after the annealing process. The Schottky metals and AlGaN/GaN heterostructure interacted actively during high temperature annealing so that the atoms in Schottky metals and AlGaN/GaN heterosturcture diffused and the surface roughness of Schottky metal was increased. Metals used for Schottky contact are attributed to these results. The threshold voltage and the leakage current of fabricated AlGaN/GaN HEMT were shifted by 0.2 V and decreased by 40 % after the annealing process. The Ni/Au Schottky contact is not stable under the 300 °C condition. More researche for the improvement of the thermal stability of Schottky contact for AlGaN/GaN based devices are required.

ACKNOWLEDGMENTS

This research was supported by Electronic Power Industry Technology R&D program through the Power IT Agency funded by the Ministry of Knowledge Economy.

REFERENCES

1. S. J. Pearton, J. C. Zolper, R. J. Shul and F. Ren, "GaN: Processing, defects, and devices", *J. Appl. Phys.*, vol. 86, no. 1, pp 1-78, July, 1999.
2. D. Ueda, T. Murata, M. Hikita, S. Nakazawa, M. Kuroda, H. Ishida, M. Yanagihara, K. Inoue, T. Ueda, Y. Uemoto, T. Tanaka and T. Egawa, "AlGaN/GaN Devices for Future Power Switching Systems", *Int. Electron Device Meeting Tech. Dig.*, pp 389-392, 2005.
3. J. W. P. Hsu, M. J. Manfra, R. J. Molnar, B. Heying, and J. S. Speck, "Direct imaging of reverse-bias leakage through pure screw dislocations in GaN films grown by molecular beam epitaxy on GaN templates", *Appl. Phys. Lett.*, vol. 81, no. 1, pp 79-81, July, 2002.
4. J. Lee, D. Liu, H. Kim and W. Lu, "Postprocessing annealing effects on direct current and microwave performance of AlGaN/GaN high electron mobility transistors", *Appl. Phys. Lett.*, vol. 85, no. 13, pp. 2631-2633, September, 2004.
5. H. Kim, M. Schuette, H. Jung, J. Song, J.Lee, Wu Lu and J. C. Mabon, "Passivation effects in Ni/AlGaN/GaN Schottky diodes by annealing", *Appl. Phys. Lett.*, vol. 89, no. 13, 053516, 2006
6. S.-C. Lee, J. Lim, M.-W. Ha, J.-C. Her, C.-M. Yun, and M.-K. Han, "High Performance AlGaNGaN HEMT Switches Employing 500°C Oxidized NiAu Gate for Very Low Leakage Current and Improvement of Uniformity", *Proc. Int. Symp. Power Semiconductor Device and ICs*, 2006, pp. 247-250.
7. M.-W. Ha, S.-C. Lee, J.-H. Park, J.-C. Her, K.-S. Seo, and M.-K. Han, "Silicon Dioxide Passivation of AlGaNGaN HEMTs for High Breakdown Voltage", *Int. Symp. Power Semiconductor Device and ICs*, 2006, pp. 169-172.
8. B. M. Green, K. K. Chu, E. M. Chumbes, J. A. Smart, J. R. Shealy, and L. F. Eastman, "The effect of surface passivation on the microwave characteristics of undoped AlGaN/GaN HEMTs", *IEEE Electron Device Lett.*, vol. 21, no. 6, pp. 268-270, June, 2000.

Compound Semiconductors for Sensing

Mater. Res. Soc. Symp. Proc. Vol. 1167 © 2009 Materials Research Society 1167-O06-03

A III-Nitride Layered Barrier Structure for Hyperspectral Imaging Applications

L. D. Bell[1], N. Tripathi[2], J. R. Grandusky[2], V. Jindal[2], and F. Shahedipour-Sandvik[2]
[1]Jet Propulsion Laboratory, California Institute of Technology, Pasadena, CA 91109
[2]College of Nanoscale Science and Engineering, University at Albany, Albany, NY 12203

ABSTRACT

We report on a novel photodetector structure based on III-nitride materials. A layered configuration is used to create a barrier with voltage-tunable height. The barrier is used as a filter for photoexcited holes and electrons, and could form the basis for a dynamically tunable pixel in a hyperspectral imaging array. This would eliminate the need for external gratings and filters used in conventional hyperspectral instruments. In addition, the tunability of pixels allows a decrease of the array dimension by one. The III-nitride materials family is a good candidate for this device, combining large band offsets with the ability for epitaxial growth. We have demonstrated the feasibility of using III-nitride materials to fabricate layered tunnel barriers and have demonstrated tunability of photodetection using these structures. External quantum efficiencies of > 12% have been achieved with prototype devices.

INTRODUCTION

Improvement in III-nitride growth methods and material quality has led to many applications of this material system for optical and electronic devices, including light emitting diodes[1,2,3,4] (LEDs), laser diodes[5] (LDs), visible and ultraviolet detectors[6], and high electron mobility transistors[7]. A proposed new application of this material system is for a layered barrier heterostructure to be used as the critical element in a tunable hyperspectral detector.

Hyperspectral imaging, or imaging spectrometry, has important applications to the remote investigation of both the Earth's surface and of the outer planets and moons [8,9] via chemical composition mapping and analysis. The characterization of surfaces and atmospheres requires chemical composition mapping and analysis, and hyperspectral imaging provides critical information on the surface composition, atmosphere, and temperature. Human-made objects can also be identified and classified. Since the technique provides chemical information on vegetation as well as inorganic materials, it is also valuable for resource identification. Deployment can be airborne or orbital.

Conventional hyperspectral imagers use an external grating to disperse the incoming light along one spatial dimension of a detector array. Different configurations for the imager are possible. A 1D detector array can be used to capture spectral information for a single spatial pixel, in conjunction with a mechanical raster to assemble a spatial image line. Vehicle motion along the flight direction provides the raster in the other spatial dimension. Thus this type of "whiskbroom" operation requires only a linear array. Alternatively, in the "pushbroom" configuration, a 2D array can be used to record spectral information and one spatial dimension, eliminating the need for the mechanical raster; vehicle motion is still used for the second spatial dimension.

This paper describes a new method for producing a detector pixel with detection sensitivity that can be tuned via application of a voltage. This wavelength-tunable pixel element is based on layered or graded barrier materials, and can be used in a dynamically tunable hyperspectral imaging detector array. Each pixel would be tunable through a range of wavelengths determined by pixel design. All spectral information for a spatial pixel could be acquired within a single physical detector pixel, allowing the detector array dimensions to be dedicated to only spatial information. This eliminates the need for external gratings and filters, substantially decreasing weight, size, and complexity and increasing robustness.

For a given mode of operation, the detector dimension could be decreased by one: "pushbroom" operation could be accomplished with a line array, and "whiskbroom" imaging would require only a single pixel detector. This reduced requirement on array dimension would lead to reduced size and complexity and greatly increased array uniformity.

This type of tunable pixel has the potential to provide a wide range of detection wavelengths, extending from near-IR to the UV region, without use of filters and gratings which make existing hyperspectral detectors cumbersome. The small size, along with the possibility of growth of the device structure on silicon substrates, enables integration with Si CMOS.

The photodetector is based on voltage-tunable electron barriers fabricated from layered dielectric materials. Peaked tunnel barriers were first proposed by Likharev [10] in the context of flash memory, to replace the conventional tunnel barrier between channel and floating gate in those devices. The important concept is that shape-engineering of the barrier potential profile enables the barrier height to be tuned by adjustment of a voltage. For a conventional square barrier, an applied voltage decreases the average height but not (to first order) the maximum height. In contrast, the maximum height of a barrier that is peaked in the center is reduced by an applied voltage (Fig. 1). For a symmetric triangular barrier, $\Delta V/V$ is 0.5. A symmetric layered barrier can also approach this value if enough layers are used.

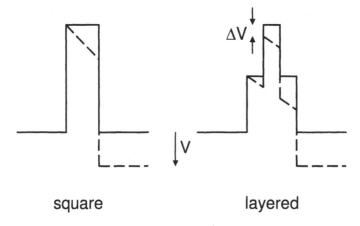

square   **layered**

Figure 1. Simplified energy band diagrams for square and layered barriers with no applied bias (solid lines) and under bias V (dashed lines), showing the barrier lowering ΔV that occurs for a peaked barrier.

For electron tunneling, this height reduction leads to a faster change in barrier conductance for a given bias voltage change, leading to faster write/erase operations in a floating-gate memory. The tunable detector described here also takes advantage of the voltage-tunable barrier height. In this case, we propose using a thicker barrier such that tunneling of equilibrium electrons is not allowed; only hot electrons with energies in excess of the barrier maximum may cross the barrier. Therefore, this height tuning provides energy selectivity for collection of photoexcited electrons and thus a spectroscopy of photon wavelength.

EXPERIMENT

The hyperspectral detector reported here utilizes a 5-layered epitaxially grown heterostructure of $Al_xGa_{1-x}N$ that approximates a triangular barrier. It has previously been reported that a triangular barrier exhibits the largest barrier lowering with applied voltage.[10] However, a true triangular barrier requires grading of the Al composition from 0% to 100% within a small thickness, which represents extremely difficult growth challenges. Therefore, GaN and AlN alloys have been used to construct a layered barrier with a stepped potential profile to approximate a triangular barrier. Height of the barrier can be tuned with applied bias, and a five-layer structure yields nearly the same magnitude of lowering as that of a triangular potential. Earlier reports on barrier lowering have also used dielectric layers with various band gaps in order to approximate a triangular barrier.[11]

The properties of the III-nitride material system make it promising for this application. The band gap can be varied from 0.65 eV (InN) to 6.04 eV (AlN)[12], alloy fraction can be graded to obtain a true triangular barrier, and it can be grown epitaxially to limit leakage current through grain boundaries. Moreover, the pseudomorphic nature of III-nitride barriers is expected to yield higher quantum efficiency (QE) than amorphous barriers. Additionally the device could be operated with injection of photoexcited carriers from the metal contact to extend the detection further into the infrared region of the spectrum, providing a versatile detector.

There are several design and growth considerations for barrier composition and layer thicknesses, based on achieving optimum performance. There should be a large difference in barrier height between the edges and the center of the barrier; this maximizes the window of wavelength tunability. Also, the structure design should minimize the contribution of different photoabsorption mechanisms in the same wavelength range; such interference between mechanisms makes deconvolution of spectral content difficult.

The barrier should be nearly opaque to equilibrium carriers near the Fermi level; this requirement corresponds to a constraint on dark current in the detector. Excessive dark current not only introduces a large unwanted current component but also limits the voltage that can be applied to the structure (limiting tunability). Thus the layered barrier must be thick enough to maintain low dark current. At the same time, the barrier should be as transmissive as possible for hot electrons at the top of the barrier; this maximizes the quantum efficiency of the detector. If the barrier becomes too thick, scattering losses become large. High-quality growth must be maintained in order to minimize defects, which is necessary to reduce dark current and also to improve QE by reducing scattering. High material quality also minimizes undesirable charging effects.

Device structures were grown by metalorganic chemical vapor deposition (MOCVD) as described previously.[13] Prior to growth of the barrier heterostructures, thick $Al_xGa_{1-x}N$ layers were grown to calculate the growth rate for different Al compositions. The device structure consists of 2 μm of n-type GaN, followed by a 5-layer barrier heterostructure of $Al_{0.33}Ga_{0.66}N/Al_{0.66}Ga_{0.33}N/AlN/Al_{0.66}Ga_{0.33}N/Al_{0.33}Ga_{0.66}N$. A 10 nm GaN layer was grown on the final AlGaN layer to serve as a contact layer. A Schottky contact of Au was deposited on top of the GaN layer for current-voltage (I-V) and internal photoemission (IPE) measurements.

SimWindows Semiconductor Device Simulator Version 1.5 was used to calculate the band profiles of the detector structures, using band offset values reported by G. Martin et al.[14]. It was also used to obtain information on the expected barrier lowering under voltage. Devices were designed with different barrier thickness values. The 5-layer structures that were grown consisted of $Al_{0.33}Ga_{0.66}N/Al_{0.66}Ga_{0.33}N/AlN/Al_{0.66}Ga_{0.33}N/Al_{0.33}Ga_{0.66}N$ layers, ranging in total thickness from 18 nm to 90 nm. The thinner barriers were designed to be in close proximity to the critical thickness values for an $Al_xGa_{1-x}N/GaN$ heterostructure calculated by Floro et al.[15] and Lee et al.[16]. The $Al_xGa_{1-x}N/GaN$ heterostructure is expected to begin strain relaxation by generation of surface cracks[16], which is not observed in atomic force microscopy (AFM) images of any of these structures.[13]

Internal photoemission (IPE), which has been used extensively to measure potential barrier profiles [17,18,19], was used to measure the barrier height and barrier lowering provided by the heterostructures. In IPE, tunable monochromatic light is incident on the sample structure. The resulting spectra of photocurrent versus wavelength exhibit structure due to the onset of hot carrier transmission across the barrier. Since both hot electrons and hot holes are generated, thresholds are observed corresponding to electron and hole transport across the conduction-band and valence-band barriers, respectively. Hot carriers of both types are generated in both the underlying GaN template layer and in the top metal layer; thus, the general transport picture includes carriers crossing the barrier in both directions. In order to simplify the picture and select a subset of these transport processes, a voltage can be applied to the barrier structure to enhance current in one direction, as shown in Fig. 2. This diagram illustrates the situation with a negative bias applied to the gate, in which hole transport from semiconductor to metal, and electron transport from metal to semiconductor, are enhanced. These appear as photocurrent contributions (of the same sign) in two different wavelength regions of the spectrum.

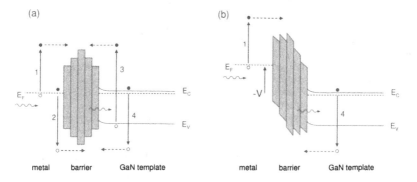

Figure 2. Schematic conduction and valence band diagram for a layered barrier structure (a) with no applied bias and (b) with negative gate bias. The applied voltage preferentially enhances certain transport processes.

RESULTS AND DISCUSSION

In order to evaluate the material quality and estimate the suitability of the layers for a tunable hyperspectral detector, IV measurements were always done prior to IPE characterization. One requirement of these structures is that they have low dark current, even with large applied bias; this is verified by IV data. The barrier structures generally yielded less than 100 nA of current at 1 V bias, although current values as much as two orders of magnitude smaller than this were observed. Due to the large (1.5 mm diameter) contact areas required for IPE, spatial variation of defects in the barrier layers could cause variation in sample resistance.

IPE measurements were performed to evaluate barrier height and device QE, as well as to study the behavior of the barrier with applied voltage. For these devices, external QE is defined as the number of electrons out per incident photon, and is calculated using $I_e/P_{ph}*E_{ph}$, where I_e is measured electron current in amps, P_{ph} is measured incident light power in watts, and E_{ph} is the

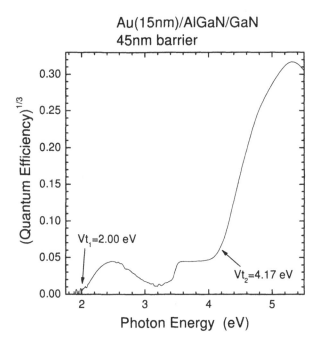

Figure 3: IPE spectrum (external quantum efficiency vs. photon energy) for a 5-layer (45 nm total thickness) barrier at 0V applied bias. Electrons from the metal produce the current contribution at low energy (~2 eV) and holes from the GaN produce the current at higher energy (~4.2 eV). The threshold at 3.4 eV is attributed to leakage of band-edge holes through barrier defects.

incident photon energy in eV. An IPE spectrum for a 45 nm barrier is shown in Fig. 3, for an applied bias voltage of 0V. Features due to carriers originating in the semiconductor are assumed to have a $(E-E_{th})^3$ dependence, where E_{th} is the threshold photon energy of the current onset, whereas features due to carriers originating in the metal are modeled with a $(E-E_{th})^2$ dependence.[19] Three main features are visible in the spectrum, corresponding to different photoexcitation and transport processes.

The threshold at 4.2 eV is due to absorption in the GaN template and transport of the photoexcited holes across the valence-band barrier into the top metal layer. This threshold value corresponds to the GaN band gap plus the additional energy needed to surmount the barrier. The value agrees well with the sum of the GaN band gap (3.4 eV) and the measured valence-band offset between GaN and AlN [14] of 0.7 ± 0.24 eV. The data in Fig. 3 is shown as the cube root of QE in order to linearize the region above this threshold, since the expected behavior is QE ~ $(E-E_{th})^3$ as mentioned above. Thus a straight-line fit of the linear region yields the threshold value. The maximum external QE of this process is about 3% in this sample; other sample structures exhibited QE of greater than 12%.

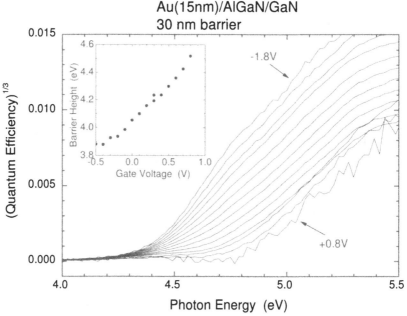

Figure 4: IPE spectra for a 5-layer (30 nm total thickness) barrier over a range of applied biases from -1.8 V to +0.8 V, with spectra displayed every 0.2 V. The current corresponds to photogenerated holes crossing the valence-band maximum of the barrier. The inset shows the variation of potential barrier height as a function of applied voltage calculated from the thresholds of the spectra, referenced to the bulk GaN conduction band.

The threshold at 2.0 eV is likely due to absorption in the metal layer and hot electron transport across the conduction-band barrier into the GaN. For these spectral features due to carriers originating in the metal layer, thresholds are extracted assuming quadratic dependence on energy, as mentioned above. The threshold value for this process is the energy difference between the top of the barrier and the Fermi level. From [14], the conduction-band offset between GaN and AlN is ~2.1 ± 0.24 eV. The Schottky barrier height between Au and clean GaN is about 1 eV [20], but can vary depending on surface conditions. Modeling indicates that the measured threshold is about 0.4 eV lower than calculated, but contributions from other factors such as polarization charge are not included in the modeling.

Finally, the threshold at 3.4 eV is due to band-gap absorption in the GaN followed by leakage of band-edge carriers through defects in the barrier. This contribution to photocurrent should not be visible in the absence of leakage paths across the barrier, since such carriers have insufficient energy to surmount the barrier.

The sign of photocurrent in Fig. 3 corresponds to electrons moving from metal to the GaN template layer, indicating that there is a built-in potential in the barrier structure that favors transport in this direction. This is in agreement with results of modeling. It was found that the applied gate voltage required to achieve flat-band conditions varied for different structures, but was usually between 0 V and +1 V; the sign was always positive.

The measured photoemission current for a range of applied bias as a function of incident photon energy is shown in Fig. 4 for a 30 nm barrier structure. The plot shows a significant decrease in the threshold energy with increasing negative applied bias. The noise level for larger positive voltages is due to the increasing leakage current at those voltages (this is the forward-bias direction for these structures). The potential barrier for photogenerated carriers can be calculated from the IPE spectrum [19,21]. The inset to Fig. 4 shows the extracted valence-band potential barrier (referenced to the GaN bulk conduction-band edge) as a function of applied bias, calculated from the thresholds of the spectra. A decrease in the potential barrier height φ_b with decreasing applied bias V_g in the range from -0.4 V to +0.8 V is clearly observed, with a ratio $\Delta\varphi_b/e\Delta V_g \approx 0.5$. This confirms that the barrier height and hence the detection wavelength range of the device can be tuned by changing the applied voltage.

CONCLUSIONS

In summary, a layered barrier structure with applications for tunable hyperspectral detection has been developed using the III-nitride material system. The device consists of an epitaxially grown 5-step heterostructure of $Al_xGa_{1-x}N/AlN/Al_xGa_{1-x}N$ that approximates a triangular barrier. Current-voltage measurements of the device structures show sufficiently low leakage currents for tunable detection to be demonstrated. Improvements in material quality through reduction of dislocation density should allow for improved forward and reverse characteristics and improved sensitivity and signal-to-noise of the devices. IPE measurements confirm a substantial decrease in barrier height with variation in applied bias. These devices also exhibit a high external QE measured without surface anti-reflection coatings and without optimization of barrier profile for maximum QE.

ACKNOWLEDGMENTS

The research described in this paper was carried out in part at the Jet Propulsion Laboratory, California Institute of Technology, under a contract with National Aeronautics and Space Administration.

REFERENCES

1. A.J. Fischer, A.A. Allerman, M.H. Crawford, K.H.A. Bogart, S.R. Lee, R.J. Kapler, W.W. Chow, S.R. Kurtz, K.W. Fullmer, J.J. Figiel, *Appl. Phys. Lett.* **84**, 3394 (2004).
2. K.S. Ramaiah, Y.K. Su, S.J. Chang, B. Kerr, H.P. Liu, I.G. Chen, *Appl. Phys. Lett.* **84**, 3307 (2004).
3. G.Y. Zhang, Z.J. Yang, Y.Z. Tong, Z.X. Qin, X.D. Hu, Z.Z. Chen, X.M. Ding, M. Lu, Z.H. Li, T.J. Yu, L. Zhang, Z.Z. Gan, Y. Zhao, C.F. Yang, *Optical Materials* **23**, 183 (2003).
4. C.H. Liu, Y.K. Su, T.C. Wen, S.J. Chang, R.W. Chang, *J. Crystal Growth*, **254**, 336 (2003).
5. S. Figge, J. Dennemarck, G. Alexe, D. Hommel, *Mater. Res. Soc. Symp. Proc.* **831**, E11.36 (2005).
6. B. Potì, M. T. Todaro, M. C. Frassanito, A. Pomarico, A. Passaseo, M. Lomascolo, R. Cingolani, and M. De Vittorio, *Electron. Lett.* **39**, 1747 (2003).
7. M. Fieger, Y. Dikme, F. Jessen, H. Kalish, A. Noculak, A. Szymakowski, P. Gemmern, B. Faure, C. Richtarch, F. Letertre, M. Heuken, and R. H. Jensen, *Phys. Stat. Sol. (c)* **2**, 2607 (2005).
8. J. S. Pearlman, P. S. Barry, C. C. Segal, J. Shepanski, D. Beiso, and S. L. Carman, *IEEE Trans. Geosci. Rem. Sens.* **41**, 1160 (2003).
9. R. Marion, R. Michel, and C. Faye, *IEEE Trans. Geosci. Rem. Sens.* **42**, 854 (2004).
10. K. K. Likharev, *Appl. Phys. Lett.* **73**, 2137 (1998).
11. J.D. Casperson, L.D. Bell and H.A. Atwater, *J. Appl. Phys.* **92**, 261 (2002).
12. T. Fujii, K. Shimomoto, R. Ohba, Y. Toyoshima, K. Horiba, J. Ohta, H. Fujioka, M. Oshima, S. Ueda, H. Yoshikawa, and Keisuke Kobayashi, *Applied Physics Express* **2**, 011002 (2009).
13. N. Tripathi, J. R. Grandusky, V. Jindal, F. Shahedipour-Sandvik, and L. D. Bell, *Appl. Phys. Lett.* **90**, 231103 (2007).
14. G. Martin, A. Botchkarev, A. Rockett and H. Morkoc, *Appl. Phys. Lett.* **68**, 2541 (1996).
15. A. Floro, D. M. Follstaedt, P. Provencio, S. J. Hearne, and S. R. Lee, *J. Appl. Phys.* **96**, 7087 (2004).
16. S. R. Lee, D. D. Koleske, K. C. Cross, J. A. Floro, K. E. Waldrip, A. T. Wise, and S. Mahajan, *Appl. Phys. Lett.* **85**, 6164 (2004).
17. J. C. Brewer, R. J. Walters, L. D. Bell, D. B. Farmer, R. G. Gordon and H. A. Atwater, *Appl. Phys. Lett.* **85**, 4133 (2004).
18. L. S. Yu, Q. J. Xing, D. Qiao, S. S. Lau, K. S. Boutros and J. M. Redwing, *Appl. Phys. Lett.* **73**, 3917 (1998).
19. R. J. Powell, *J. Appl. Phys.* **41**, 2424 (1970).
20. L. D. Bell, R. P. Smith, B. T. McDermott, E. R. Gertner, R. Pittman, R. L. Pierson, and G. J. Sullivan, *Appl. Phys. Lett.* **76**, 1725 (2000).
21. E. O. Kane, *Phys. Rev.* **127**, 131 (1962).

Materials Growth and Characterization

Mater. Res. Soc. Symp. Proc. Vol. 1167 © 2009 Materials Research Society 1167-O07-04

Crystalline Perfection of Epitaxial Structure: Correlation With Composition, Thickness, and Elastic Strain of Epitaxial Layers

Balakrishnam R Jampana[1], Nikolai N Faleev[2], Ian T Ferguson[3], Robert L Opila[1], Christiana B Honsberg[4]
[1]Material Science and Engineering, University of Delaware, Newark, Delaware 19716, USA
[2]Electrical and Computer Engineering, University of Delaware, Newark, Delaware 19716, USA
[3]School of Electrical and Computer Engineering, Georgia Institute of Technology, Atlanta, Georgia 30332, USA
[4]Electrical Engineering, Arizona State University, Tempe, Arizona 85287, USA

ABSTRACT

Crystalline perfection of InGaN epi-layers is the missing design parameter for InGaN solar cells. Structural deterioration of InGaN epi-layers depends on the thickness, composition and growth conditions as well. Increasing the InGaN epi-layer thickness beyond a critical point introduces extended crystalline defects that hinder the optical absorption and electrical properties. Increasing the InGaN composition further reduces this critical layer thickness. The optical absorption band edge is sharp for III-nitride direct band gap materials. The band edge profile is deteriorated by creation of extended crystalline defects in the InGaN epitaxial material. The design of InGaN solar cells requires the growth of epi-layers where a trade off between crystalline perfection and optical absorption properties is reached.

INTRODUCTION

InGaN alloys with band gap variability in the 0.7 to 3.4 eV are used commercially to make blue and UV LEDs and lasers. This band gap range makes it possible to develop multiple junction solar cells to span most of the solar spectrum and achieve ultra high efficiencies [1, 2]. Wide band gap (> 2.2 eV) InGaN solar cells are being currently explored, owing to the band gap limitation of other materials. InGaN homo-junction solar cells with 2.0 V open circuit voltage have been demonstrated under AM1.5 illumination [3], but the short circuit current in these solar cells is low. InGaN p-i-n solar cells and quantum well solar cells have also been demonstrated [4, 5] but in most cases the open circuit voltage or the short circuit currents are low. The degradation of all these solar cells can be correlated back to the crystalline quality of the InGaN epitaxial material.

InGaN epitaxial material has been studied for almost two decades now. Most of the research was focused on development of LEDs and lasers. The thickness of the active layer is less than 10 nm in these device structures. The solar cells require thicker InGaN material (> 100 nm) determined by the absorption coefficients. The issues originating with growth of thicker InGaN material and their effect on device performance have not been addressed extensively in the past research. In this paper, we present the spatial distribution of crystalline defects in InGaN epitaxial layers and correlate these with the observed optical absorption properties.

The paper presents the observed crystalline quality results for $In_{0.12}Ga_{0.88}N$ epi-layers with increasing thickness followed by results with varying InGaN (13.5, 16 and 20%) composition. These results are followed by the observed optical absorption properties. The interpretation of the observed crystalline defects and their influence on optical absorption properties are then presented.

GROWTH and CHARACTERIZATION

Epitaxial Growth

Thin undoped InGaN layers were grown on standard undoped-GaN/sapphire templates in an Emcore MOCVD reactor. Trimethylgallium (TMG), Triethylgallium (TEG) and Trimethylindium (TMI) were the precursors used to introduce gallium and indium, respectively, into the reactor using hydrogen as the carrier gas; while ammonia (NH_3) was used as the nitrogen source. The undoped -GaN layer was grown using a two step process, where typically a 20-40 nm thin GaN buffer layer growth at 550°C is followed by the epitaxy of high-quality 2 μm thick GaN template at 1030°C. To study the effect of growth conditions on crystalline quality of InGaN epitaxial layers two sets of samples were grown. The first set consisted of four $In_{0.12}Ga_{0.88}N$ samples with 50, 100, 200 and 400 nm thicknesses, grown on the template at 780°C. The growth rate of GaN is set at 2.15 μm/hr, while that of $In_{0.12}Ga_{0.88}N$ is 0.116 μm/hr. The second set consisted of three 200 nm thick InGaN samples with 13.5%, 16% and 20% InGaN composition. The growth rate of GaN is set at 1.9 μm/hr, while that of InGaN at 0.23 μm/hr with growth temperature in the 650 to 680 °C range.

Material Characterization

High-resolution X-ray Diffraction (HRXRD) studies were performed with an X'Pert diffractometer featuring Ge(220) four bounce monochromator with $CuK_{\alpha 1}$ monochromatic radiation and Ge(220) three bounce analyzer in triple-axis alignment. Reciprocal Space Maps (RSMs) of symmetrical (0002) and asymmetrical (11.4) reflections were collected to specify the extent of relaxation of elastic stress and the type and spatial distribution of crystalline defects. The samples were characterized for optical reflection and transmission using PerkinElmer Lambda 1050 UV/Vis/NIR system in the 250 to 700 nm range.

RESULTS

The crystalline quality was analyzed collectively for two cases: one considering the variation in thickness of 12% InGaN composition samples and the other by considering 200 nm fixed thickness with varying InGaN composition.

High resolution ω-2θ triple-axis rocking curves for samples with varying thickness are shown in Figure 1(a). The full width half maximum (FWHM) of the GaN and InGaN peaks along with fringes around the InGaN peak are a guideline to determine the crystalline nature of the epitaxial layers. The rocking curves with 50 and 100 nm InGaN thickness have fringes matching to the simulated curves (not included here), while these fringes are washed out in the 200 and 400 nm thick InGaN samples. The asymmetric reciprocal space maps (ARSM) for 50 and 200 nm InGaN samples are shown in Figure 2, with markings showing the stress relaxation. Stress relaxation is negligible ($\Delta Q_x=0$) for InGaN epi-layers thicknesses upto 100 nm, beyond which observable relaxation ($\Delta Q_x \neq 0$) begins in 200 nm thick InGaN.

Similar high resolution ω-2θ rocking curves for samples with varying InGaN composition are shown in Figure 1(b). The FWHM increases with increasing InGaN composition and no fringes around the peaks are observed. The ARSM for 16% InGaN and 20% InGaN compositions are shown in Figure 3. The extent of relaxation from ARSM is calculated to be ~5, 10 – 12 and 16 – 18% for 13.5, 16 and 20% InGaN compositions, respectively.

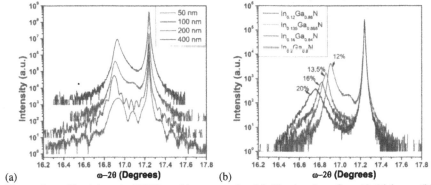

(a) ω–2θ **(Degrees)**   (b) ω–2θ **(Degrees)**

Figure 1. ω-2θ triple-axis (0002) rocking curve for (a) Varying $In_{0.12}Ga_{0.88}N$ thickness (b) Varying InGaN composition with fixed 200nm thickness

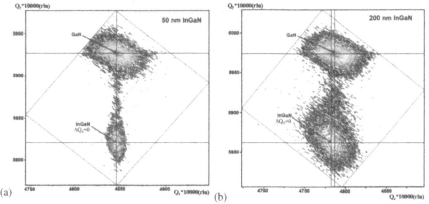

(a)   (b)

Figure 2. Asymmetric (11.4) RSM for $In_{0.12}Ga_{0.88}N$. Blue lines and red lines intersect at GaN and InGaN peak positions, respectively. (a) 50 nm thick (ΔQ_x=0, strained) (b) 200 nm thick (ΔQ_x≠0, partial relaxation)

The extent of stress relaxation, vertical coherence length and density of threading dislocations are quantitatively extracted from HRXRD. The vertical coherence length and density of threading dislocation are graphed in Figure 4 for varying thickness of 12% InGaN composition. The vertical coherence length in GaN decreases with increasing InGaN thickness and is pinned to around 470 nm beyond 200 nm InGaN thickness. The vertical coherence length for InGaN increases with InGaN thickness and is pinned to around 115 nm beyond 200 nm InGaN thickness. The density of threading dislocations increases in both GaN and InGaN with increasing InGaN thickness but is more drastic in the thicker InGaN epi-layer, after the process of structural transformation of point defects was shifted to the InGaN layer. Similar results are tabulated for varying InGaN composition in Table 1. The density of threading dislocations increased with InGaN composition.

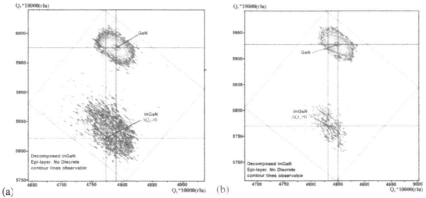

Figure 3. Asymmetric (11.4) RSM for 200 nm thick (a) 16% InGaN ($\Delta Q_x \neq 0$, partial relaxation), (b) 20% InGaN compositions ($\Delta Q_x \neq 0$, partial relaxation). Blue lines and red lines intersect at GaN and InGaN peak positions, respectively.

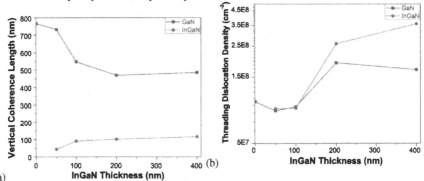

Figure 4. Calculated crystalline quality parameters from ω-2θ rocking curves (a) vertical coherence length (b) density of threading dislocations as a function of $In_{0.12}Ga_{0.88}N$ thickness

The optical reflection and transmission on samples were measured and the absorption characteristics were calculated and shown in Figure 5. The increase in thickness of InGaN with fixed 12% InGaN composition results in a sharp optical absorption edge up to a thickness of 100 nm, beyond which absorption below the band gap is observed, shown in Figure 5(a). In the case with varying InGaN composition samples, see Figure 5(b), the expected sharp absorption edge has completely deteriorated.

Table 1. Determined crystalline quality parameters for varying InGaN composition

	13.5% InGaN	16% InGaN	20% InGaN
Density of Threading Dislocations (cm^{-2}) (in InGaN)	3.15×10^8	1.9×10^9	3.75×10^9
Extent of Relaxation	5%	20%	30%

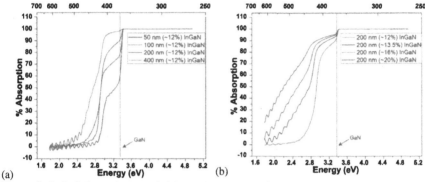

(a)   (b)

Figure 5. Optical absorption recorded for (a) Varying In$_{0.12}$Ga$_{0.88}$N thickness (b) Varying InGaN composition with fixed 200nm thickness

DISCUSSION

InGaN epi-layer thickness is an important parameter in the design of solar cells. A trade off among the design constraints on thickness is the crystalline perfection, optical absorption and carrier diffusion. The crystalline perfection of the In$_{0.12}$Ga$_{0.88}$N epi-layer degrades beyond 100 nm. As the thickness of InGaN is increased beyond this "critical" thickness, extended crystalline defects such as dislocation loops are formed in the volume of the InGaN epi-layer. The InGaN epi-layer can now be interpreted as two sub-layers: the top sub-layer has less crystalline defects compared to the bottom transformation sub-layer, with both sub-layers having the same InGaN composition. The approximate demarcation between the two-sub layers is determined from the vertical coherence length and the thickness of InGaN determined from growth. The thickness of the less crystalline defects sub-layer is equal to the vertical coherence length and the difference between the total epi-layer thickness and less crystalline defects sub-layer gives the bottom sub-layer thickness. The formation of the extended crystalline defects in the volume of the epi-layer may be explained by creation, diffusion, accumulation and structural transformation of point defects during the epitaxial growth [6, 7, 8]. The ARSM indicate the lack of observable relaxation of elastic stress in the InGaN epi-layers up to 100 nm thickness, beyond which noticeable relaxation is observed, caused by bottom edge segments of dislocation loops and possibly misfit dislocations at the GaN/InGaN interface. These dislocations hinder the vertical carrier transport and hence reduce the carrier collection efficiency of the solar cells. The corresponding optical absorption of the samples indicates a sharp absorption edge up to a thickness of 100 nm, beyond which absorption below the band gap is observed. This is most likely related to the edge segments of the secondary dislocation loops, created in the volume of the InGaN epi-layer (at the top of the transformation sub-layer). The absorption below the band gap is an issue in the overall design of multiple junction solar cells. Ideally, all energy below the band gap has to be transmitted. However, with thickness beyond 100 nm, this would pose concerns for multiple junction solar cells.

The InGaN epi-layers with increasing (13.5, 16 and 20%) InGaN compositions have a larger extent of relaxation in comparison with the same thickness of 12% InGaN. The multi sub-layer structure discussed above is also formed in these samples. The "critical" InGaN layer thickness is less than 100 nm due to the definite surface layer decomposition. The optical absorption band edge degrades drastically with increasing composition related to creation of more extended crystalline defects in the InGaN epi-layers. The increase in density of threading dislocations by an order of magnitude is an indication of the InGaN epi-layer decomposition with increasing InGaN composition.

CONCLUSIONS

The issue limiting efficient InGaN solar cells is crystalline perfection of InGaN epi-layers. The optimal thickness of InGaN epi-layer determined by absorption coefficients is a few hundred nano-meters, but the crystalline quality deteriorates with increasing InGaN thickness, and more rapidly with increasing composition. Optical absorption below the band gap of the material is observed for thicknesses beyond the "critical" thickness and related to the creation of specific extended crystalline defects in the volume of the InGaN epi-layer. The creation of extended crystalline defects further increases with increasing composition. Also, increasing extent of relaxation of elastic stress hinders the vertical carrier transport in the material. The design of functional InGaN solar cells requires the growth of epi-layers where the trade off between crystalline perfection and optical absorption properties is balanced.

REFERENCES

1. E. Trybus, G. Namkoong, W. Henderson, S. Burnham, W. A. Doolittle, M. Cheung, and A. Cartwright, Journal of Crystal Growth, vol. 288, pp. 218-224, 2006
2. O. Jani, I. Ferguson, C. Honsberg, and S. Kurtz, Applied Physics Letters, vol. 91, p. 132117, 2007
3. Omkar K Jani, Balakirshnam Jampana, Mohit Mehta, Hongbo Yu, Ian T Ferguson, Robert Opila, Christiana B Honsberg, Proceedings of 33rd IEEE PVSC, San Diego, May 2008
4. R. Dahal, B. Pantha, J. Li, J. Y. Lin, H. X. Jiang, Applied Physics Letters, v 94, n 6, p 063505, Feb. 2009
5. C.J. Neufeld, N.G. Toledo, S.C. Cruz, M. Iza, S.P. DenBaars, U.K. Mishra, Applied Physics Letters, v 93, n 14, p 143502, Oct. 2008
6. Nikolai Faleev, Balakrishnam Jampana, Anup Pancholi, Omkar Jani, Hongbo Yu, Ian Ferguson, Valeria Stoleru, Robert Opila, and Christiana Honsberg, Proceedings of 33rd IEEE PVSC, San Diego, May 2008
7. N. Faleev, C. Honsberg, O. Jani, and I. Ferguson, Journal of Crystal Growth, vol. 300, pp. 246-50, 2007
8. N. Faleev, H. Lu, W.J. Schaff, Journal of Applied Physics, 101, 093516 (2007)

Mater. Res. Soc. Symp. Proc. Vol. 1167 © 2009 Materials Research Society   1167-O07-05

Investigation of Composition-Dependent Optical Phonon Modes in $Al_xGa_{1-x}N$ Epitaxial Layers Grown on Sapphire Substrates

Jun-Rong Chen, Tien-Chang Lu, Hao-Chung Kuo, and Shing-Chung Wang

Department of Photonics and Institute of Electro-Optical Engineering, National Chiao Tung University, 1001 University Rd., Hsinchu, 300 Taiwan

ABSTRACT

We reported the systematical study of optical properties of hexagonal $Al_xGa_{1-x}N$ epitaxial films grown on c-sapphire substrate using metal-organic chemical vapor deposition. By performing Fourier transform infrared spectroscopy measurements, the high-frequency dielectric constants and phonon frequencies can be obtained by theoretically fitting the experimental infrared reflectance spectra using a four-phase layered model. The high-frequency dielectric constant of $Al_xGa_{1-x}N$ varies between 4.98 and 4.52 for $\varepsilon_{\infty,\perp}$ (polarization perpendicular to the optical axis) and between 4.95 and 4.50 for $\varepsilon_{\infty,//}$ (polarization parallel to the optical axis) respectively when the aluminum composition changes from 0.15 to 0.24. Furthermore, from experimental infrared reflectance spectra of $Al_xGa_{1-x}N$ films, a specific absorption dip at 785 cm^{-1} was observed when the aluminum composition is larger than 0.24. The dip intensity increases and the dip frequency shifts from 785 to 812 cm^{-1} as aluminum composition increases from 0.24 to 0.58. According to the reciprocal space map of x-ray diffraction measurements, the emergence of this dip could be resulted from the effects of strain relaxation in AlGaN epitaxial layers due to the large lattice mismatch between GaN and AlGaN epitaxial film.

INTRODUCTION

Hexagonal GaN and AlN semiconductors, and AlGaN alloys, have attracted considerable attention due to their successful applications in the fabrication of high-performance electronic and optoelectronic devices. By varying the alloy composition, different electrical and optical properties can be obtained in a wide spectral range from 3.4 to 6.3 eV. In order to further engineer these alloys and related optoelectronic devices, it is necessary to work on the fundamental properties of these materials. The infrared optical response of these alloys is important for the determination of crystal quality and phonon properties. Recently, Sun et al. proposed the idea of developing terahertz quantum cascade lasers with GaN/AlGaN quantum-well structures, which have more advantages than GaAs-based material system [1]. Yu et al. studied the infrared reflectivity spectra of GaN and $Al_xGa_{1-x}N$ with aluminum compositions of 0.087, 0.27, and 0.35 [2]. They found that the E_2 mode, which arises from the disordered state of the alloys, can be observed in the refractivity spectrum of $Al_xGa_{1-x}N$. Holtz et al. reported optical studies on $Al_xGa_{1-x}N$ alloy layers grown on (111)-oriented silicon substrates by combining Fourier transform infrared (FTIR) and Raman spectroscopy studies [3]. Besides, Hu et al. studied the optical properties of hexagonal $Al_xGa_{1-x}N$ (x from 0.05 to 0.42) epitaxial films with Si doping concentration up to 10^{18} cm^{-3} grown on c-plane sapphire substrates using infrared reflectance spectra [4]. The longitudinal-optical phonon plasmon (LPP) coupled modes of n-type hexagonal $Al_xGa_{1-x}N$ films were also discussed in their studies. Nevertheless, the phonon-related studies on $Al_xGa_{1-x}N$ epitaxial films mostly focus on the composition-dependent infrared spectroscopy. To the best of our knowledge, the effects of strain relaxation of $Al_xGa_{1-x}N$ epitaxial films on infrared reflectance spectra are hardly reported in literature.

In this study, we will discuss the composition dependence of infrared optical phonon modes in $Al_xGa_{1-x}N$ epitaxial layers grown on c-plane sapphire substrates using Fourier transform infrared reflectance measurements. The effects of strain relaxation of $Al_xGa_{1-x}N$ epitaxial films on infrared reflectance spectra are investigated as well. Furthermore, the infrared reflectance spectra of GaN/sapphire and $Al_xGa_{1-x}N$/GaN/sapphire are theoretically fitted by using the factorized form of the dielectric function for the entire frequency region measured.

EXPERIMENT AND THEORY

Hexagonal $Al_xGa_{1-x}N$ epitaxial films in an entire aluminum composition range were grown on c-plane sapphire substrates by metal-organic chemical vapor deposition (MOCVD). AlGaN films were deposited using a low-pressure vertical reactor (EMCORE D75) with a fast rotational susceptor. For all samples, a normal 30-nm-thick GaN nucleation layer was deposited at 500 °C. Before growing a 0.4-μm-thick $Al_xGa_{1-x}N$ film, a 2-μm-thick GaN buffer layer was grown on the GaN nucleation layer. All the growth conditions of GaN buffer layers were the same. $Al_xGa_{1-x}N$ epilayers in composition range of $0 < x < 0.3$ were grown in N_2 and H_2 mixture ambient gas and at pressure of 100 Torr. For the composition range of $0.3 < x < 1$, the $Al_xGa_{1-x}N$ epilayers were grown in N_2 and H_2 mixture ambient gas and at pressure of 50 Torr in order to obtain a better aluminum incorporation efficiency [5]. For the AlN epilayers, in order to obtain better crystal quality, the samples were grown in a pure N_2 ambient gas and at pressure of 100 Torr. The incident angle with respect to the plane of incidence was kept at 75° (Brewster's angle was about 68°). The infrared reflectance spectrum of one-side-polished c-sapphire substrate was measured as well and no mathematical smoothing has been performed for the experimental reflectance spectra. The sample size should be lager than 1.5×1.5 cm^2 to collect the all reflected beams.

Since hexagonal $Al_xGa_{1-x}N$ epilayers are uniaxial crystals, the p (electric-filed vector E parallel to the plane of incidence) and s (electric-field vector E perpendicular to the plane of incidence) modes of plane electromagnetic waves can not be independent of each other as the incident beam is oblique with optical axis, which is c-axis in hexagonal crystal system. Therefore, we employ a 4×4 matrix in order to calculate optical response of anisotropic media. This method is provided by Schubert and can be applied in the well-known transfer matrix method for multi-layer anisotropic media [6]. The dielectric functions of anisotropic films in the x–y plane (perpendicular to the c axis, $\varepsilon_x = \varepsilon_y$) and in the z plane (parallel to the c axis) are $\varepsilon_\perp$ and $\varepsilon_{//}$, respectively. It is noteworthy that the hexagonal nitride-based materials are uniaxial crystals. The s-polarized mode, whose electric-field vector E is perpendicular to the optics axis is known as the ordinary rays. Therefore, the s-polarized reflectance spectra have similar spectral shape under different angles of incidence. However, the p-polarized mode, whose electric-field vector E is perpendicular to the ordinary rays is the extraordinary rays. The p-polarized reflectance spectra are sensitive to the angles of incidence, which can effectively response the optical anisotropic properties. Since the reststrahlen regions of sapphire overlap with those of GaN and AlGaN, it is difficult to identify the phonon frequencies directly from the measured infrared reflectance spectra. Separately modeling the reflectance spectra of sapphire, GaN/sapphire, and

AlGaN/GaN/sapphire is necessary to extract the phonon modes and dielectric function of each layer. A three-phase layered model is used to calculate the reflectance spectra of air/GaN/sapphire. We applied the obtained fitting parameters of the three-phase model in the subsequent four-phase model calculations for the $Al_xGa_{1-x}N$ epilayers (air/Al_xGa_{1-x}N/GaN/sapphire) [7]. For polar semiconductor materials, the contribution of l polar crystal lattice modes to the infrared dielectric function at photon energy $\hbar\omega$ can be expressed using a factorized model with Lorentzian broadening [2, 8-10],

$$\varepsilon(\omega) = \varepsilon_{\infty,j} \prod_{i=1}^{l} \frac{\omega_{LO,ij}^2 - i\gamma_{LO,ij}\omega - \omega^2}{\omega_{TO,ij}^2 - i\gamma_{TO,ij}\omega - \omega^2},$$ (1)

where j is equal to "//" or "⊥", which denotes the dielectric functions and phonon frequencies parallel or perpendicular to the optic c axis, respectively. $\omega_{LO,ij}$, $\gamma_{LO,ij}$, $\omega_{TO,ij}$, and $\gamma_{TO,ij}$ represent the phonon frequency and the broadening value of the ith LO and TO phonon, respectively. The parameters $\omega_{TO,\perp}$, $\omega_{TO,//}$, $\omega_{LO,\perp}$, and $\omega_{LO,\perp}$ are equal to the common used notations of the frequencies of the $E_1(TO)$, $A_1(TO)$, $E_1(LO)$, and $A_1(LO)$ modes, respectively. The model parameters $\varepsilon_{\infty,//}$ and $\varepsilon_{\infty,\perp}$ are the high-frequency dielectric constants for polarization parallel and perpendicular to the optical axis, respectively. The best-fit parameter values including high-frequency dielectric constants and phonon modes (ω_{TO} and ω_{LO}) in equation (1) can be obtained using a Levenberg-Marquardt algorithm [11].

RESULTS AND DISCUSSION

In order to investigate the phonon frequencies of GaN and $Al_xGa_{1-x}N$ epilayers on sapphire substrates, it is necessary to obtain the knowledge of the dielectric function for sapphire substrate and identify the influences from sapphire phonon modes. The infrared dielectric anisotropy and phonon modes of sapphire are completely studied by Schubert et al. using infrared spectroscopic ellipsometry [10]. They determined the ordinary and extraordinary infrared complex dielectric functions as well as all infrared-active phonon modes of single crystal c-sapphire for wavelengths from 3 to 30 μm. In this study, we directly applied all physical parameters of c-sapphire including two high-frequency dielectric constants, four distinct infrared-active modes with dipole-moment oscillation perpendicular to the c axis, and two modes with dipole-moment oscillation parallel to the c axis from the reported values by Schubert et al. The experimental (solid lines) and theoretical (dashed lines) infrared reflectance spectra of sapphire, GaN, AlN, and $Al_xGa_{1-x}N$ epitaxial films with aluminum composition x from 0.15 to 0.58 are shown in figure 1. It is obvious that there is a sharp dip at 755 cm^{-1} in GaN and all $Al_xGa_{1-x}N$ spectra compared with the spectra of sapphire. This dip is induced by LO phonon due to the optical anisotropy in the GaN material. Specifically, the p-polarized light of nonperpendicular incidence in the hexagonal epilayers can be interacted with the LO phonon mode and make it infrared active. For experimental reflectance spectra of $Al_xGa_{1-x}N$ films, the incorporation of 15% aluminum results in an additional dip at about 648 cm^{-1} as compared with that of GaN. Moreover, the dip intensity increases with the dip position slight shift of toward higher frequency when the aluminum composition increases in the range from 0.15 to 0.58. The dip was also observed in the experimental reflectance spectra of $Al_xGa_{1-x}N$ measured by Yu et al. [2]. They supposed the origin of this dip is related to the E_2 mode, which results from the random distribution of alloy constituents and the elimination of the translational symmetry of the lattice.

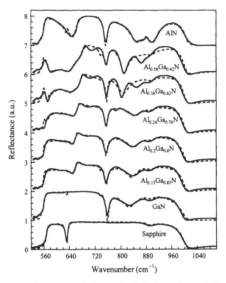

Figure 1. Experimental (solid lines) and theoretical fitting (dashed lines) infrared reflectance spectra of sapphire, GaN on sapphire, and $Al_xGa_{1-x}N$ films on sapphire with different aluminum compositions.

In figure 1, it is found that when the aluminum composition is smaller than 0.24, the theoretical fitting reflectance spectra according to equation (1) are in good agreement with experimental results. However, when the aluminum composition is 0.24, a new dip at about 785 cm^{-1} can be observed in the experimental reflectance spectrum. The dip intensity increases with increasing aluminum composition and the dip frequency shifts from 785 to 812 cm^{-1} as aluminum composition increases from 0.24 to 0.58. In order to further investigate the origin of this dip, the reciprocal space maps (RSMs) of x-ray diffraction intensity of $Al_{0.2}Ga_{0.8}N$ and $Al_{0.24}Ga_{0.76}N$ films were performed around an asymmetrical GaN (10$\bar{1}$5) Bragg peak as shown in figure 2. As depicted in figure 2 (a), the maximum of the $Al_{0.2}Ga_{0.8}N$ reciprocal lattice points is at fully strained position. Since the layer thicknesses of all $Al_xGa_{1-x}N$ films are nearly the same, we assume that the lattice constant of $Al_xGa_{1-x}N$ films with aluminum composition smaller than 0.2 is fully coherent to that of GaN bulk. Furthermore, when the aluminum composition of $Al_xGa_{1-x}N$ films increases from 0.2 to 0.24, the maximum of the $Al_{0.24}Ga_{0.76}N$ reciprocal lattice points shifts from a fully strained to a partially relaxed position, as shown in figure 2 (b). The degree of strain relaxation shall progressively increase with increasing aluminum composition under the condition of similar AlGaN layer thickness due to the large lattice mismatch between GaN and AlN. By comparing figures 1 and 2, it is found that the emergence of the dip at 785 cm^{-1} and the strain relaxation of AlGaN film occur at the same aluminum composition. Moreover, the dip intensity increases with aluminum composition, which results in the increased degree of strain relaxation as well. Therefore, we deduce that the origin of this dip is attributed to the strain relaxation of $Al_xGa_{1-x}N$ films. Since the phonon properties are closely related to the lattice vibration, crystal structure, and alloy composition, it can be expected that the fully strained, partially relaxed, and fully relaxed AlGaN films shall eliminate the translational symmetry of the lattice, which influences the phonon properties and is characterized from the infrared reflectance spectra. It is reasonable that the measured infrared reflectance spectra of $Al_xGa_{1-x}N$ films with aluminum composition larger than 0.24 cannot be excellently fitted by the symmetric physical model. The effect of strain relaxation of $Al_xGa_{1-x}N$ films should be considered in the determination of phonon frequency since it is difficult to grow fully strained AlGaN film with high aluminum composition.

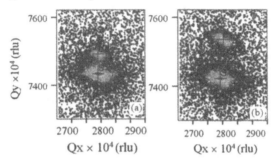

Figure 2. Reciprocal space maps of (a) $Al_{0.2}Ga_{0.8}N$ and (b) $Al_{0.24}Ga_{0.76}N$ epitaxial films.

The best fitting parameters of GaN/sapphire, and $Al_xGa_{1-x}N$/GaN/sapphire with aluminum composition from 0.15 to 0.24 are shown in Table I. Since the fitting parameters of the reflectance spectra of $Al_xGa_{1-x}N$ films with aluminum composition larger than 0.24 are not reliable due to the mismatch between measured and calculated reflectance spectra, these parameters are not shown in Table II. The anisotropic high-frequency dielectric constants $\varepsilon_{\infty,\perp}$ and $\varepsilon_{\infty,//}$ decrease and the phonon frequency of ω_{TO} and ω_{LO} increase with increasing aluminum composition. The high-frequency dielectric constants of $Al_xGa_{1-x}N$ decrease from 4.98 to 4.52, which are located between the values of 5.01 for $Al_{0.087}Ga_{0.913}N$ and 4.5 for $Al_{0.27}Ga_{0.73}N$ [2].

Table. I. Best fitting parameters of GaN/sapphire as well as $Al_xGa_{1-x}N$/GaN/sapphire with aluminum composition from 0.15 to 0.24 determined by infrared reflectance spectra.

	GaN	$Al_{0.15}Ga_{0.85}N$	$Al_{0.2}Ga_{0.8}N$	$Al_{0.24}Ga_{0.76}N$
$\varepsilon_{\infty,\perp}$	5.11	4.98	4.77	4.52
$\varepsilon_{\infty,//}$	5.07	4.95	4.65	4.50
$\omega_{TO,\perp}$ (cm^{-1})	559.1	565.2	570.3	575.8
$\omega_{TO,//}$ (cm^{-1})	533.1	542.1	545.3	550.7
$\omega_{LO,\perp}$ (cm^{-1})	742.2	775.1	787.6	790.2
$\omega_{LO,//}$ (cm^{-1})	734.5	762.1	775.8	780.6
$\omega_{TO,\perp}$ (cm^{-1})		628.6	625.9	623.7

CONCLUSIONS

In summary, we have grown AlGaN films with different aluminum compositions for the study of composition-dependent phonon mode energies. The phonon mode frequency can be obtained from theoretically fitting the experimental FTIR spectra. However, we found that the calculated reflectivity spectra cannot be excellently fitted by the physical model when the aluminum composition is larger than 24%. According to asymmetrical RSMs measured around the GaN $(10\bar{1}5)$ reflection, we suggest that this condition attributed to the emergence of a new dip at 785 cm^{-1} which can be observed as the strain relaxation of AlGaN film occurred. An advanced physical model is necessary for fitting the infrared reflectivity spectra of relaxed AlGaN films.

ACKNOWLEDGMENTS

This work is supported by the MOE ATU program and, in part, by the National Science Council of the Republic of China under Contract Nos. NSC 96-2221-E009-092-MY3, NSC 96-2221-E009-093-MY3, NSC 96-2221-E009-094-MY3, and US Air Force Research Laboratory.

REFERENCES

1. G. Sun, R. A. Soref, and J. B. Khurgin, Superlattice. Microstruct. 37, 107 (2005).
2. G. Yu, H. Ishikawa, M. Umeno, T. Egawa, J. Watanabe, T. Soga, and T. Jimbo, Appl. Phys. Lett. 73, 1472 (1998).
3. M. Holtz, T. Prokofyeva, M. Seon, K. Copeland, J. Vanbuskirk, S. Williams, S. A. Nikishin, V. Tretyakov, and H. Temkin, J. Appl. Phys. 89, 7977 (2001).
4. Z. G. Hu, M. Strassburg, N. Dietz, A. G. U. Perera, A. Asghar and I. T. Ferguson, Phys. Rev. B 72, 245326 (2005).
5. G. S. Huang, H. H. Yao, T. C. Lu, H. C. Kuo, and S. C. Wang, J. Appl. Phys. 99, 104901 (2006).
6. M. Schubert, Phys. Rev. B 53, 4265 (1996).
7. Z. G. Hu, M. Strassburg, A. Weerasekara, N. Dietz, A. G. U. Perera, M. H. Kane, A. Asghar, and I. T. Ferguson, Appl. Phys. Lett. 88, 061914 (2006).
8. D. W. Berreman and F. C. Unterwald, Phys. Rev. 174, 791 (1968).
9. A. Kasic, M. Schubert, S. Einfeldt, D. Hommel, and T. E. Tiwald, Phys. Rev. B 62, 7365 (2000).
10. M. Schubert, T. E. Tiwald, and C. M. Herzinger, Phys. Rev. B 61, 8187 (2000).
11. W. H. Press, S. A. Teukolsky, W. T. Vetterling, and B. P. Flannery, Numerical Recipes in C: The Art of Scientific Computing, Cambridge, Cambridge University Press, MA, 1992.

Mater. Res. Soc. Symp. Proc. Vol. 1167 © 2009 Materials Research Society   1167-O07-07

Hydrogen -related defects in bulk ZnO

Matthew D. McCluskey,[1] Slade J. Jokela,[1] and Marianne C. Tarun[1]

[1]Washington State University, Pullman, WA 99164-2814

ABSTRACT

Zinc oxide (ZnO) has attracted resurgent interest as an active material for energy-efficient lighting applications. An optically transparent crystal, ZnO emits light in the blue-to-UV region of the spectrum. The efficiency of the emission is higher than more "conventional" materials such as GaN, making ZnO a strong candidate for solid-state white lighting. Despite its advantages, however, ZnO suffers from a major drawback: as grown, it contains a relatively high level of donors. These unwanted defects compensate acceptors or donate free electrons to the conduction band, thereby keeping the Fermi level in the upper half of the band gap. This paper reviews recent work on hydrogen donors and nitrogen-hydrogen complexes in ZnO.

INTRODUCTION

Zinc oxide (ZnO) is an 'old' semiconductor that has attracted resurgent interest as an electronic material for numerous applications [1]. A wide-bandgap ($E_g \sim 3.4$ eV) semiconductor, ZnO emits light in the blue-to-UV region of the spectrum. The high efficiency of the emission [2] makes ZnO a strong candidate for solid-state white lighting. ZnO is currently used as a transparent conductor [3] in solar cells [4]. It is also a UV-absorbing material in sunscreens [5] and the active material in varistors [6]. ZnO is preferred material for transparent transistors, devices that could be incorporated in products such as liquid-crystal displays [7].

ZnO has practical benefits that make it an attractive material from an industrial point of view. In contrast to GaN, large single crystals can be grown relatively easily [8]. The environmental impact and toxicity are lower than most other semiconductors - ZnO is actually used as a dietary supplement in animal feed [9]. The low cost of zinc as compared to indium make it economically competitive for transparent conductor applications.

For ZnO to reach its potential, however, the role of donor defects must be understood. One such defect is hydrogen, a common impurity in ZnO. We have studied hydrogen donors using infrared (IR) spectroscopy and Hall-effect measurements [10,11]. One donor species results in an IR absorption peak corresponding to O-H bond-stretching vibrations. This species is unstable, decaying in a few weeks at room temperature. By correlating the free-electron concentration with the IR absorption strength, we established that hydrogen acts as a shallow donor.

Novel schemes for introducing acceptor dopants must be investigated if ZnO-based devices are to become technologically and economically feasible. Toward that end, we formed nitrogen-hydrogen (N-H) complexes in ZnO during chemical vapor transport (CVT) growth, using ammonia as an ambient [12]. The N-H bond-stretching mode gives rise to an IR absorption peak at 3150.6 cm^{-1}. Isotopic substitutions for hydrogen and nitrogen result in the expected frequency shifts, thereby providing an unambiguous identification of these complexes. The N-H complexes are stable up to ~700°C. The introduction of neutral N-H complexes could prove useful in achieving reliable p-type conductivity in ZnO.

Some of the information presented here will appear in an upcoming Focus Review article in Journal of Applied Physics.

EXPERIMENTAL DETAILS

IR spectra were obtained using a Bomem DA8 Fourier transform IR spectrometer, with a spectral resolution of 1-4 cm^{-1}. A liquid-nitrogen cooled InSb detector was used to collect the light. Low temperature (9 K) spectra were obtained with a Janis closed-cycle helium cryostat with zinc selenide windows. Hall-effect measurements (MMR, Inc.) were taken in the van der Pauw geometry with indium contacts.

The hydrogen donor work was performed using commercial single-crystal bulk samples from Cermet, Inc. For the nitrogen-hydrogen studies, polycrystalline and single-crystal ZnO samples were grown using chemical vapor transport. ZnO powder was placed at one end of a sealed ampoule, along with graphite powder, in an ammonia (NH$_3$) ambient. The ampoule was placed in a three-zone Mellen furnace, which provided a temperature gradient. Details are given in Ref. [13]. For single-crystal growth, a c-cut Cermet sample was attached to the cold end of the ampoule with a copper weld. The ZnO grew on the Cermet seed to a thickness of 1-2 mm.

DISCUSSION

Interstitial hydrogen donors

By measuring the pressure and polarization dependence of the O-H vibrational mode, and comparing to the predictions of first-principles calculations [14], Jokela and McCluskey [15] attributed the 3326 cm^{-1} peak to hydrogen donors in the antibonding (AB$_\perp$) configuration [16]. The pressure dependence for various defect-hydrogen complexes in ZnO were calculated as well [17].

However, muon-implantation experiments [18] and ab initio calculations [19] indicate that hydrogen donors should reside in the bond-centered (BC$_{//}$) configuration, not the AB$_\perp$ configuration. Experiments by Lavrov et al. [19] on hydrothermally grown ZnO samples revealed a hydrogen peak at 3611 cm^{-1} at liquid-helium temperatures, not the 3326 cm^{-1} peak that we observed. They attributed this peak to hydrogen in the BC$_{//}$ configuration. Seager and Myers observed a similar peak, although shifted and broadened, at room temperatures [20]. Work by Shi et al. [21] showed that samples annealed in hydrogen generally show the 3611 cm^{-1} and 3326 cm^{-1} peaks, but in different ratios depending on the type of sample. For example, Eagle-Picher samples show a strong 3611 cm^{-1} peak whereas Cermet samples show a strong 3326 cm^{-1} peak.

This sample-dependent variation suggests that one or both of the peaks correspond to hydrogen complexes. Results from secondary ion mass spectroscopy (SIMS) show significant concentrations of Al in samples of bulk single-crystal ZnO obtained from Cermet Inc., Ga and B in samples from Eagle-Picher, and Si in both [22]. The SIMS results showed that Cermet samples contained significant concentration of Ca impurities, raising the possibility that the O-H donors that we observed are, in fact, complexed with Ca. First-principles calculations suggest that Ca impurities indeed trap hydrogen donors and may explain the 3326 cm^{-1} peak [23].

Nitrogen -hydrogen co mplexes

A possible route toward p-type conductivity involves the incorporation of neutral N-H complexes, followed by annealing to drive out hydrogen. Jokela and McCluskey [24] reported N-H complexes in polycrystalline ZnO grown by chemical vapor transport (CVT) in an ammonia ambient. The N-H bond-stretching mode gives rise to an IR absorption peak at 3150.6 cm^{-1}. Isotopic substitutions of deuterium for hydrogen and ^{15}N for ^{14}N resulted in the appropriate frequency shifts. The N-H complexes were stable up to ~700°C. First-principles calculations that predicted that hydrogen attaches to a nitrogen atom, nearly perpendicular to the c axis [25]. The vibrational modes predicted by the calculations were in good agreement with the experimental observations of Jokela and McCluskey. Calculations show that the N-H bond is more stable than the O-H bond for interstitial hydrogen, also in agreement with experiment [26].

Recently, we incorporated N-H complexes in single crystal ZnO, enabling us to perform polarized IR spectroscopy to determine the orientation of the N-H dipole. The results for IR light propagating perpendicular to the c axis are shown in Fig. 1. For light polarized along the c axis (0°), the N-H absorption peak intensity is minimized. For light polarized along perpendicular to the c axis (90°), the absorption is maximized. These results are in agreement with the structure predicted from first principles calculations [25], shown by the ball-and-stick model.

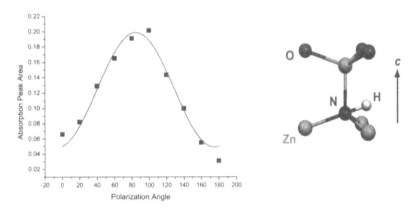

Figure 1. Absorption of the N-H stretch mode peak versus polarization angle (relative to the c axis). The ball-and-stick model is the proposed structure for N-H complexes in ZnO.

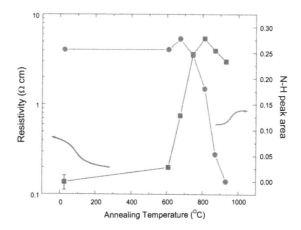

Figure 2. Electrical resistivity and area of the N-H IR peak, as a function of temperature for 1-hr isochronal anneals in air. The O-H peak was not present in the sample.

The single-crystal ZnO:N,H sample was annealed in the air for 1-hr durations. The results of these isochronal anneals are shown in Fig. 2. The resistivity increases for annealing temperatures greater than 600°C, consistent with the out-diffusion of hydrogen donors. For temperatures above 700°C, the N-H complex begins to dissociate. As grown, the sample had a carrier concentration of $n \sim 5 \times 10^{17}$ cm^{-3} and a mobility of $\mu \sim 100$ cm^2/Vs. After the anneal at 745°C, the values were $n \sim 3 \times 10^{16}$ cm^{-3} and $\mu \sim 60$ cm^2/Vs. It is not clear why the resistivity drops for higher temperatures. It is possible that contaminants such as In diffused into the ZnO and contributed to the n-type conductivity.

Interestingly, when polycrystalline samples are grown with small quantities of Mg and Ca in the ampoule, in an argon ambient, we observe the N-H peak. This may be due to a reaction between Mg (or Ca) and unintentional N_2 contaminants. We are currently investigating this phenomenon with isotopically enriched N_2 introduced into the growth.

CONCLUSIONS

If defects in ZnO can be controlled, it will emerge as an important material for energy applications. Hydrogen passivates nitrogen acceptors in ZnO, and also acts as a shallow donor. Our results on annealing N-H complexes suggest that it may be possible to achieve p-type conductivity through a suitable annealing schedule.

ACKNOWLE DGMENTS

This work was supported by the U.S. National Science Foundation under Grant No. DMR-0704163 and the U.S. Department of Energy under Grant No. DE-FG02-07ER46386.

REFERE NCES

1. S.J. Pearton, D.P. Norton, K. Ip, Y.W. Heo, and T. Steiner, Journ. Vacuum Sci. Tech B 22, 932 (2004).
2. D.C. Look, Mater. Sci. Engin. B 80, 383 (2001).
3. T. Minami, MRS Bulletin 25 (8), 38 (2000).
4. A. Nuruddin and J.R. Abelson, Thin Solid Films 394, 49 (2001).
5. G.P. Dransfield, Radiation Protection Dosimetry 91, 271 (2000).
6. D.R. Clarke, Journal of the American Ceramic Society 82, 485 (1999).
7. J.F. Wager, Science 300, 1245 (2003).
8. J.M. Ntep, S.S. Hassani, A. Lusson, A. Tromson-Carli, D. Ballutaud, G. Didier, and R. Triboulet, Journ. Crystal Growth 207, 30 (1999).
9. J.W. Smith et al., Journal of Animal Science 75, 1861 (1997).
10. M.D. McCluskey, S.J. Jokela, K.K. Zhuravlev, P.J. Simpson, and K.G. Lynn, Appl. Phys. Lett. 81, 3807 (2002).
11. S.J. Jokela and M.D. McCluskey, Phys. Rev. B 72, 113201 (2005).
12. S.J. Jokela and M.D. McCluskey, Phys. Rev. B 76, 193201 (2007).
13. S.J. Jokela, M.D. McCluskey, and K.G. Lynn, Physica B 340-342, 221 (2003).
14. S. Limpijumnong and S.B. Zhang, Appl. Phys. Lett. 86, 151910 (2005).
15. S.J. Jokela and M.D. McCluskey, Phys. Rev. B 72, 113201 (2005).
16. C.G. Van de Walle, Phys. Rev. Lett. 85, 1012 (2000).
17. M.G. Wardle, J.P. Goss, and P.R. Briddon, Appl. Phys. Lett. 88, 261906 (2006).
18. K. Shimomura, K. Nishiyama, and R. Kadono, Phys. Rev. Lett. 89, 255505 (2002).
19. E.V. Lavrov, J. Weber, F. Börrnert, C.G. Van de Walle, R. Helbig, Phys. Rev. B 66, 165205 (2002).
20. C.H. Seager and S.M. Myers, J. Appl. Phys. 94, 2888 (2003).
21. G.A. Shi, M. Stavola, S.J. Pearton, M. Thieme, E.V. Lavrov, and J. Weber, Phys. Rev. B 72, 195211 (2005).
22. M.D. McCluskey and S.J. Jokela, Physica B 401-2, 355 (2007).
23. X.B. Li, S. Limpijumnong, W.Q. Tian, H.B. Sun, and S.B. Zhang, Phys. Rev. B 78, 113203 (2008).
24. S.J. Jokela and M.D. McCluskey, Phys. Rev. B 76, 193201 (2007).
25. X. Li, B. Keyes, S. Asher, S.B. Zhang, S.-H. Wei, T.J. Coutts, S. Limpijumnong, and C.G. Van de Walle, Appl. Phys. Lett. 86, 122107 (2005).
26. J. Hu, H.Y. He, and B.C. Pan, J. Appl. Phys. 103, 113706 (2008).

.

Mater. Res. Soc. Symp. Proc. Vol. 1167 © 2009 Materials Research Society   1167-O07-09

ZnO Thin Film Deposition on Sapphire Substrates by Cemical Vapor Deposition

Zhuo Chen[1], T. Salagaj[2], C. Jensen[2], K. Strobl[2], Mim Nakarmi[1], and Kai Shum[1, a]

[1]Physics Department, Brooklyn College of the City University of New York, 2900 Bedford Avenue, Brooklyn, NY 11210, USA

[2]First Nano, A Division of CVD Equipment Cooperation, 1860 Smithtown Avenue, Ronkonkoma, NY 11779, USA

[a]kshum@brooklyn.cuny.edu

ABSTRACT

ZnO thin films with thickness around 200 nm were deposited on a-plane sapphire substrates by Chemical Vapor Deposition (CVD) method with a mixed ZnO-powder/C-powder solid source. These films were characterized by Atomic Force Microscopy (AFM), Scanning Electron Microscopy (SEM), and photoluminescence (PL) spectroscopy. The correlation between surface structural properties of ZnO thin films and their optical signature measured by temperature dependence of PL is investigated for various growth conditions such as flow rate O_2 injection gas and growth temperature. At room temperature, the columbic interaction enhanced absorption edge of 3.305 eV of these films was determined by optical absorption measurements.

INTRODUCTION

ZnO is very interesting and promising optical material because it has a room temperature direct band gap of 3.37 eV [1] and excitonic binding energy of 60 meV [2]. However, it remains very challenging to make it as a truly useful optoelectronic material. One of the challenges is to epitaxially grow ZnO on a suitable substrate although thin film deposition on various substrates such as silicon or sapphire wafers using different methods has been successful [3]. Another challenge is to understand and control the high level of non-intentionally doped residual electron density. This not-yet controllable residual electron density [4, 5] is the reason behind the inability of making ZnO a good p-type material necessary optoelectronic devices. It also limits the full potential of large exciton binding energy because of carrier screening effect.

In this work, we report on the experimental data resulting from our attempt to understand how ZnO can be grown on a sapphire substrate by a solid-vapor-solid process in a commercially built chemical vapor deposition (CVD) apparatus using a solid source. Various growth parameters such as growth temperature and O_2 gas injection rate were used to alter the ZnO nucleation process. Deposited films were characterized by scanning electron microscopy and photoluminescence. We found that ZnO films could be grown by the CVD method with the film quality comparable to these grown by more expensive apparatuses such as metal-organic chemical vapor deposition (MOCVD).

EXPERIMENTAL DETAILS

A commercially built vapor deposition (CVD) system (First Nano, ET2000) was used to deposit thin film on sapphire substrates. It was equipped with a separated solid source heater, a three-zone (load, center, and end zone) furnace, a gas injector, a vacuum pump, and a quartz tube. The reaction tube was controlled by a three-zone furnace to obtain a uniform temperature across the substrate at the collecting area over 3 inches by 2 inches. The source material was mixed ZnO powder (Alfa Aesar, 99.99%) and graphite powder (Alfa Aesar, 99%) with mass ratio of 1:4. The solid source was placed at the load zone in the quartz tube and heated up to high temperature by an additional solid source heater to generate Zn vapor which was then carried into the center zone by the Ar carrying gas. The reacting gas (O_2) was introduced into the system by gas injector at different locations to achieve the best uniformity over a given substrate size. Sapphire substrates with a-plane cut were used. Fig. 1a shows the schematic diagram for the CVD apparatus. Fig 1b displays typically measured temperature profiles across load to center zone. Two important parameters, growth (substrate) temperature (T_g) and O_2 flow rate (F_{O2}) will be discussed in this work. The surface morphology of deposited thin films was studied by scanning electron microscopy (SEM) and atomic force microscopy (AFM). Optical transmission/absorption spectra were measured from PerkinElmer Lambda950 optical spectrometer. Photoluminescence (PL) spectra were measured by Horiba NanoLog system coupled with an optical cryostat (4 to 350 K) from Advanced Research System.

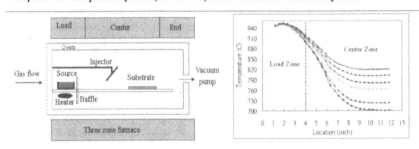

Fig. 1a   Fig.1b

RESULTS AND DISCUSIONS

Fig. 2a, 2b, and 2c display the SEM surface images for the 3 samples placed in the center zone with different growth temperature of 720, 820, and 920 °C, respectively, for 1 hour deposition time. It is clear that at 720 °C ZnO starts to nucleate on the sapphire surface, but the speed the film growth is too slow, leading to the scattered ZnO islands on the sapphire surface during 1 hour time period. At 820 °C, the speed of nucleation has dramatically increased and left a thin layer of ~ 200 nm during 1 hour. Since the ZnO coverage on the sapphire surface is not 100%, therefore, the thickness of ZnO thin film was measured by AFM as shown in Fig. 3. At 920 °C, no visible ZnO film was deposited on the substrate.

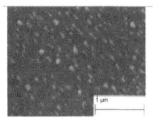  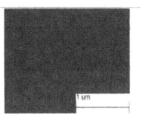

| Fig. 2a ZnO on sapphire substrate at 720 °C. | Fig. 2a ZnO on sapphire substrate at 820 °C. | Fig. 2a ZnO on sapphire substrate at 920 °C. |

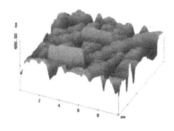

Fig. 3 AFM image of the 820 °C sample.

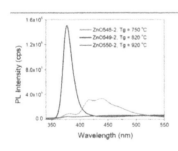 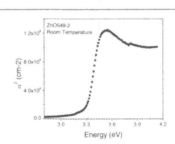

| Fig.4a PL spectra from 3 samples with different growth temperature. | Fig. 4b Tauc plot to obtain the optical band gap for the 820 °C sample. |

Photoluminescence was measured from these 3 samples at room temperature with a Xe-lamp set at 300 nm with band pass of 5 nm as a photoexcitation source. They are shown in Fig. 4a as green, blue, and red curves for the sample with the growth temperature of 720, 820, and 920 °C, respectively. All three spectra show a band edge emission at ~378 nm with negligible green emission at 550 nm, normally attributed to defect-related emission. PL from the film with growth temperature of 820 °C is certainly strongest among the three samples, consistent with the information provided by their SEM images. The origin of three weak PL peaks from 400 to 500 nm for the 720 °C sample is not clear.

In Fig. 4b, a Tauc plot [6] is displayed for the 820 °C sample. The optical band gap of 3.31 eV is obtained. This energy is consistent with other ZnO thin films obtained by other methods [7, 8] and with generally accepted ZnO band gap [1] at room temperature by taking into the excitonic effect on absorption edge.

With apparently optimal growth temperature of 820 °C on sapphire substrate, three ZnO films were deposited with different O_2 flow rate at 100, 500, and 1,000 standard cubic center meter per minute (with a short notation of cc). The SEM images for these three samples are shown in Fig. 5a, 5b, and 5c, respectively. For the 100 cc sample, it is clearly visible that there is a large amount of surface defects (pits). However, the estimated number of pit density seems to be on the order of 10^7 cm^{-2}, comparable with these thin films deposited [9] by more expensive apparatus MOCVD. Such defect density decreases as the O_2 flow rate increases as shown in the 500 cc and 1000 cc samples.

| Fig. 5a ZnO film on sapphire substrate with O2 flow rate of 100 cc. | Fig. 5b ZnO film on sapphire substrate with O2 flow rate of 500 cc. | Fig. 5c ZnO film on sapphire substrate with O2 flow rate of 1,000 cc. |

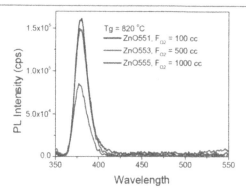

Fig. 6 PL spectra from the three samples with the different O_2 flow rate as indicated in the figure.

PL spectra for the three samples with different O_2 flow rates are displayed in Fig. 6. The PL peak intensity seems to be consistent with the pit density as obtained from SEM images. Although the data points are limited, apparently, the PL intensity (I_{PL}) has a square root relationship with the O_2 flow rate, i.e., $I_{PL} \sim F_{O2}{}^{1/2}$.

CONCLUSIONS

In summary, we have presented our preliminary data on ZnO thin film deposition by CVD method using a solid source. It was found that the optimal growth temperature is 820 °C and the film surface pit density decreases while photoluminescence intensity increases as the O_2 flow rate increases. These films clearly show good poly-crystalline quality with low optically active defect density as evidenced by the lack of so-called green emission around 550 nm.

ACKNOWLEDGMENTS

The authors would like to acknowledge the partial financial support from NYSTRA grant for this work through the CUNY center of advanced technology on photonic applications.

REFERENCES

1. E.Mollwo, in *Semiconductors: Physics of II-VI and I-VII Compounds, Semimagnetic Semiconductors*, edited by O.Madelung, M. Schulz, and H. Weiss (Springer, Berlin, 1982), vol. 17 of *Landolt-B¨ornstein New Series*, p. 35.

2. E. O'Kane, Phys. Rev. B **18**, 6849 (1978).

3. X. W. Sun and H. S. Kwok, J. Appl. Phys. **86**, 408 (1999).

4. U. Ozgur et al., J. Appl. Phys. **98**, 043101 (2005).

5. D. L. Look, J. Electronic Materials **35**, 1299 (2006).

6. J. Tauc, Amorphous and Liquid Semiconductors (Plenum, London, 1974).

7. S. W. Whangbo et al., J. of Korean Physical Society **33**, 456 (2000).

8. V. Cracium et al., Appl Phys. Lett. **65**, 2963 (1994).

9. N. Oleynik, Ph.D. Dissertation, University of Magdeburg, July 2007.

Appendix

This Appendix contains papers from
Materials Research Society Symposium Proceedings
Volume **1171E**

Materials in Photocatalysis and Photoelectrochemistry for
Environmental Applications and H_2 Generation
A. Braun, P.A. Alivisatos, E. Figgemeier,
J.A. Turner, J. Ye, E.A. Chandler, Editors

Mater. Res. Soc. Symp. Proc. Vol. 1171 © 2009 Materials Research Society   1171-S01-04

Photocatalysis Approach for Energy and Environmental Challenges at Indian Institute of Chemical Technology, Hyderabad, India

V. Durga Kumari *, M. Subrahmanyam, M.V. Phanikrishna Sharma, J. Krishna Reddy, K. Lalitha
Catalysis and Physical Chemistry Division
Indian Institute of Chemical Technology, Hyderabad 500607, India

ABSTRACT

The R & D developments in several aspects of catalysis area require cleaner and clean up technologies. Catalysts are used for energy conversion and to convert environmentally hazardous materials into harmless compounds. This presentation reviews the work currently under exploration at IICT that illustrates the perspective of photocatalyst technologies for solving energy and environmental issues for providing sustainable development. Studies on development of photocatalytic materials for degradation of phenolic wastes, common industrial effluent, H-acid, Calmagite (an azodye), Isopruturon (herbicide) and for E-coli disinfection are highlighted. Materials like Natrotantite, Ce-modified zeolites, Ag_2O/TiO_2, CuO/TiO_2 and C,N-doped TiO_2 are designed and evaluated for photocatalytic splitting of water for generation of hydrogen energy. Furthermore, potential applications of photo catalysts in the chemical synthesis of N-containing heterocyclic compounds like pyrazines and piperazines which are useful intermediates in the synthesis of various drugs, perfumes, herbicides and dyes are new interesting aspects in the presentation. Thus the present review describes the emerging trends in using photocatalysts for energy and environmental applications.

INTRODUCTION

Though TiO_2 in anatase form is the best photocatalyst reported so far, poor adsorption properties lead to great limitation in exploiting the photocatalyst to the best of it's photoefficiency. The other problem with the semiconductor photocatalyst is the faster electron-hole recombination that results in decreased photo-efficiency. Several attempts have been made to improve the photo efficiency of titania by adding adsorbents like silica, alumina, zeolites, clays and active carbon. This addition is expected to induce synergism as the pollutants are adsorbed prior to photodegradation making the process more facile. Supported TiO_2 is commonly reported to be less photoactive due to the interaction of TiO_2 with the support during the thermal treatments. A method of supporting TiO_2 on zeolites without losing the photoefficiency of TiO_2 and the adsorption properties of the support is the important aspect while preparing zeolite based photocatalysts. Synergistic effects were also evidenced when mixtures of TiO_2 and adsorbents like active carbon were employed for photoxidation. TiO_2 based zeolites are extensively prepared, characterized and evaluated in this laboratory for the degradation of phenolic waste and also for the synthesis of N-heterocyclics using both UV and solar irradiation. TiO_2 and TiO_2 supported zeolites are also drawn into thin films using acrylic emulsions as binder and applied for large scale applications for treating common industrial effluent containing dye and drug intermediates. TiO_2 is also supported on mesoporous materials like Al-MCM-41 and SBA-15 for treating pesticide containing wastewater in order to achieve higher degradation rates. Ce and Bi modified zeolites are showing photocatalytic properties in water splitting reaction and photodegradation of phenol. Whereas Fe modified zeolites are heterogenised photo-fenton catalysts used in phenol degradation. Ce modified Al-MCM-41 is also used as support to disperse TiO_2 wherein Ce ion is stabilized in lower oxidation on photoirradiation acting as electron acceptor and minimizing the electron-hole recombination. TiO_2 is also sensitized by doping metal ions as well as nonmetals to obtain visible light activity. Thus Ag_2O/TiO_2, CuO/TiO_2 and C, N-doped TiO_2 are established for water splitting to generate hydrogen using solar energy. Natrotantite which is normally synthesized in solid state at very high temperatures was attempted in hydrothermal conditions and established for structure and as a water splitting catalyst under UV irradiation. TiO_2 supported on pumice stone, Hβ and hydroxy apatite (HAP) are evaluated for disinfection of water. The materials prepared, characterized and evaluated as photo catalyst for energy and environmental applications are summarized below and their fine tuning and development for specific applications is a continuous study in this laboratory.

DISCUSSION

Photocatalysts developed for the degradation of organic wastes

Bi-ZSM-5 : This catalyst is prepared by impregnation of Bi $(NO_3)_3.5 H_2O$ over HZSM-5 zeolite. Isolated Bi_2O_3 particles are present over the support with interaction. The adsorption properties of the support and the increased band gap of interacted bismuth resulted in enhanced activity of Bi-ZSM-5 catalysts in the photodegradation of phenol under UV light [1].

Sm^{3+} doped Bi_2O_3 : Sm^{3+} doped Bi_2O_3 photo catalyst is prepared by hydrothermal process. DRS showed a large decrease in the band gap from 2.85 to 2.1 eV. The low band gap energy, high surface area and decreased rate of e^-/h^+ recombination by Sm doping increased the photocatalytic activity of the catalyst in the degradation of methylene blue under solar light. The decreasing order of the activity under solar irradiation is: Sm^{3+} doped $Bi_2O_3 > TiO_2 > Bi_2O_3$ [2].

Ce-Al-MCM-41 : Ce-Al-MCM-41 is prepared by the impregnation of Ce $(NO_3)_3.6H_2O$ over Al-MCM-41. Cerium ion in interaction with Al-MCM-41is in +3 oxidation state and showing the higher photocatalytic activity compared to pure ceria. Ce^{3+} on irradiation generates electrons that participate in effective photodegradation of phenol under UV light [3].

TiO_2/Ce-Al-MCM-41 : TiO_2/Ce-Al-MCM-41 composite is prepared by solid state dispersion of P-25 (TiO_2) over Ce-Al-MCM-41. The Ce^{4+} ion formed on photoexcitation of Ce^{3+} ion, traps the electron of TiO_2 and minimizes the e^-/h^+ recombination over titania. The synergism resulted in enhanced activity of the composite catalyst for the degradation of phenol under UV light [4].

Nano C, N doped TiO_2 : Mesoporous nano C, N doped TiO_2 is prepared by the hydrolysis of titanium isopropoxide. Pyridine is used as the source of C and N doped in to the TiO_2 lattice. Pyridine developed the selective linkage of Ti-O-N in C, N doped-TiO_2 evidenced by the IR and XPS techniques. The catalysts show high visible light activity for the degradation of 4-nitro phenol and MB under solar light [5].

TiO_2 /HZSM-5 Acrylic films : These films are made with a proprietary binder using a spray technique onto an inert support. HZSM-5 zeolite is immobilized in a stable polyacrylic film. It is an economical, practical and promising method to immobilize photocatalysts for the elimination of xenobiotic pollutants from wastewaters. TiO_2/HZSM-5 films are showing higher activity compared to TiO_2 under UV light [7].

Fe (III)-HY : Fe (III)-HY catalysts are simply prepared by impregnation method. IR, DRS and XPS studies confirm the redox states of Fe before and after the reaction. No leaching of Fe from the solid catalyst into the solution is detected by AAS. Heterogeneous Fe (III)-HY is more efficient compared to homogenous Photo-Fenton system that can be applied at pH>3 [8]. The catalyst is showing good activity with a minimum concentration of H_2O_2 and Fe for the degradation of phenol under UV light [8].

TiO_2 and TiO_2-AE film : TiO_2 slurry and TiO_2-AE film are coated on a cuddapah flat stone support and used in the degradation of H-acid. The catalyst is stable and showing enhanced activity that recommended for large scale continuous operation [9].

TiO_2 Suspension : Commercial TiO_2 samples are used in suspensions and evaluated for the degradation of orange red. P-25 suspension is showing higher activity compared to PC 50, PC 100 and PC 500 TiO_2 [10].

Photocatalysts developed for water splitting

Ag_2O/TiO_2 : This catalyst was prepared by impregnation method. XRD and TEM show fine dispersion of silver on TiO_2. XPS establishes the interaction of silver with TiO_2. DRS shows a red shift of 50 nm in the absorption band edge of titania making the catalyst visible sensitive. H_2 production about 145μmol h^{-1}g^{-1} cat is observed with pure water. In 5% aqueous methanol the H_2 production is around 3350μmol h^{-1} g^{-1} cat. Continuous H_2 production is observed due to the reversibility in the redox properties of silver ion ($Ag^+ + e^- \leftrightarrow Ag$). [11]

CuO/TiO_2 : This catalyst was prepared by impregnation method. XRD shows fine dispersion of Cu on TiO_2 surface. DRS shows a shoulder in the absorption edge of TiO_2. XPS results indicate the interaction of Cu with TiO_2. H_2 production about 581μmol h^{-1}g^{-1}cat is observed in pure water. In 5% aqueous methanol the hydrogen production observed is around 19117μmol h^{-1} g^{-1} cat. [12]

C, N-doped TiO_2 : This catalyst was prepared by hydrolysis method. A single crystalline anatase phase is observed by XRD. N_2 adsorption measurements show that catalysts are mesoporous with high surface area. Ti-O-N and Ti-C bonds are confirmed by XPS studies. C, N doped TiO_2 with oxynitride linkages are showing shift in the band gap towards the visible region. H_2 production about 700μmol h^{-1} g^{-1} cat is observed in nano TiO_2 catalyst in 10% aqueous methanol. In C,N-doped TiO_2 catalyst the rate of hydrogen production is around 3400μmol h^{-1} g^{-1} cat. Improvement in activity is observed for C, N doped TiO_2 catalyst compared to nano TiO_2 due to C doping into TiO_2 lattice and the incorporation of N preferentially into interstitial positions of the titania lattice. [13]

Natrotantite ($Na_2Ta_4O_{11}$) : This catalyst was prepared by hydrothermal synthesis. Nano crystalline natrotantite is observed in ethanol washed sample. Mesoporous and crystalline samples resulted during the high temperature calcinations. Nano and mesoporosity of the sample are confirmed by TEM and N_2-adsorption studies. H_2 production of 14.10 μmol h^{-1}g^{-1}cat is observed in pure water. In presence of methanol the H_2 production reached 83.30 μmol h^{-1}g^{-1}cat. In high temperature calcined samples, 7.40 μmol h^{-1}g^{-1}cat of H_2 was produced. This work highlights the preparation of nano crystalline Natrotantite through surfactant assisted hydrothermal synthesis at low temperatures [14].

Ce-modified Zeolites : These catalysts were prepared by impregnation method. XPS of Ce-zeolites with low Ce content shows strong interaction between zeolites and cerium and most of the cerium is in Ce^{3+} state. The rate of H_2 production over Ce- MCM-41, Ce-Al-MCM-41, Ce-HY, Ce-HZSM-5 is 1416, 651, 121 and 231μmol $h^{-1}g^{-1}$cat respectively. With increasing Al content the activity decreased as Al traps the excited electrons. , The photo irradiation of Ce^{3+} species generates electrons and forms redox couple Ce^{+3}/Ce^{+4} ($Ce^{3+} + h\nu \rightarrow Ce^{4+} + e^-$) [15].

Photocatalysts developed for pesticide degradation

TiO$_2$/HY : HY zeolite with Si/Al $\approx$ 2.6 and surface area of 340 m^2g^{-1} is used as a support. TiO$_2$ is dispersed over it by SSD method. 10 wt% loading of TiO$_2$ is found to be optimum for the degradation of isoproturon. TiO$_2$/HY shows decreased surface area of 130 m^2g^{-1}. The SEM-EDAX data proved that nearly 9.8 % TiO$_2$ is present on the surface of HY. Isoproturon is completely degraded with in 120 min with the rate of 5.95 E-06 and nearly 80% mineralization was observed with in 5h over 1 gL^{-1} catalyst amount. [16]

TiO$_2$/H-MOR : H-MOR zeolite with $SiO_2/Al_2O_3 \approx$ 20 and surface area of 440 m^2g^{-1} is used as a support. TiO$_2$ is dispersed over it by SSD method. 15 wt% TiO$_2$ loading is found to be optimum for this study and its surface area is 352 m^2g^{-1}. Isoproturon is completely degraded within 45 min with the rate of 1.16E-05 and ~87% mineralization was noted within 5h over 1.5 gL^{-1}catalyst amount [17].

TiO$_2$/AlMCM-41 : Al-MCM-41 with Si/Al $\approx$ 15.5 and surface area of 1124 m^2g^{-1} is prepared by hydrothermal synthesis using a template and all TiO$_2$ loadings over the support are prepared by SSD method. 10wt% loading was to be found optimum for this study and the catalyst is having surface area of 742 m^2g^{-1}. Isoproturon is degraded completely with in 90 min with the rate of 6.79E-06 and ~88% mineralization was observed within 5h over 1 gL^{-1}catalyst amount. [18]

TiO$_2$/PNS : Porous nano silica (PNS) material with a surface area of 153 m^2g^{-1} is prepared by salt mediated synthesis and over this material TiO$_2$ is finely dispersed by SSD method. Only 5wt% of loading found to be optimum and having surface area of 115 m^2g^{-1}. XRD of PNS is showing clearly the low angle with mesoporous range and these are confirmed by N_2 adsorption-desorption studies and TEM. Isoproturon is completely degraded with in 90min with the rate of 7.88E-06 and ~81% mineralization was occurred within 5h over 1 gL^{-1}catalyst amount. Optimum catalyst amount for this reaction is 3 gL^{-1} and its rate is 2.12E-05. Also technical Pirimicarb is degraded within 60 min in the same conditions. Commercial pesticides Imidacloprid and Phosphamidon were also checked for degradation in the same conditions and were found to degrade with in 240 and 120 mins respectively. [19]

TiO$_2$/SBA-15 : SBA-15 with a surface area of 624 m^2g^{-1} is prepared by hydrothermal synthesis and TiO$_2$ is dispersed over it by SSD method. 10 wt% loading of TiO$_2$ was found to be optimum for this reaction and its surface area is 432 m^2g^{-1}. XRD is showing clearly the low angle (100, 110 and 200 planes) with mesoporous range and it is supported by TEM and N_2 adsorption-desorption studies. Isoproturon is degraded with in 30min at a rate of 1.88E-05 and ~91% mineralization was observed with in 5h (completely mineralized in 9h) over 1 gL^{-1} catalyst amount. Technical Pirimicarb was degraded within 45 min in the same conditions. Commercial pesticides Imidacloprid and Phosphamidon were degraded under the same conditions within 180 and 60 mins respectively. Consortium of pesticides (Isoproturon 20 ppm, Imidacloprid and Phosphomidon each 15 ppm) total 50 ppm are successfully degraded within 90 min. [20]

Industrial Effluent Treatment

TiO₂ : In this study photo catalytic de-coloriztion and mineralization of common industrial effluent is obtained with TiO_2 photo catalyst using solar light illumination. Also solar photo catalytic coupled biological treatment of N-containing organic compounds in wastewater are tested successfully for pyrazinamide drug manufacturing effluent [21,22]

Photocatalytic disinfection of water

TiO₂ over pumice, Hβ and HAP : Disinfection and detoxification are illustrated in our recent work [23] using TiO_2 over pumice stone, Hβ [24] and hydroxyapatite [25]. These are taken as supports for TiO_2 and used for treatment of bacterial inactivation (E coli), which are commonly existing in real waters and also different organic pollutant degradations like acid orange-7, resorcinol, 4,6 dinitro-o-cresol, 4-nitrotoluene-2-sulfonic acid, isoproturon are performed [23] The water that was passed through the pumice stone supported reactors demonstrated strong bactericidal ability as well as detoxification affects by showing reduction in TOC, extent of decrease in color over treatment period. The photo catalytic deactivation of total E coli by solar light has been also attempted for real river waters. The technical feasibility and performance of photo catalytic effects of TiO_2 coatings over pumice stone in continuous reactor processes studied proved to improve the potability of drinking water collected from real waters.

Photocatalytic organic synthesis

TiO₂ over Zeolites : Also several N-heterocyclic compounds like dihydropyrazine (DHP), trans-1,4,6,9-tetraazabicyclic[4.4.0] decane (TAD), 2-methyl-quinoxaline (2-MQ), quinoxaline (Q), 2-methyl pyrazine (2-MP) and piperazine (P) have been synthesized in artificial UV light and solar light using TiO_2/Zeolite as photocatalyst [26-28].

CONCLUSIONS

This paper describes some of the existing photocatalyts developed in our laboratory. Although new catalysts are greatly needed in various areas, it is not easy to realize those specific catalysts due to severe requirements for the catalytic performance. The design of photo active sites, making them more visible light absorption thin film materials for large scale water treatment are some of the potential areas for the development with high efficiency and long sustainability. The technologies for the wastewater applications or organic synthesis or fuel production either using artificial or solar light, will compete in the market to develop more precise material preparation methods.

ACKNOWLEDGEMENTS

The authors thank CSIR, MOEF, MNRE, IFCPAR, MCF, DST, New Delhi for funding.

REFERENCES

[1] J. Krishna Reddy, K. Lalitha, V. Durga Kumari, M. Subrahmanyam, Catal. Lett. 121, 131 (2008)

[2] V. Durga Kumari, presented at National Seminar on Industrial Ecology ECO, Nagpur, 2008 (Unpublished)

[3] J. Krishna Reddy, V. Durga Kumari, M. Subrahmanyam, Catal. Lett. 123, 301 (2008)

[4] J. Krishna Reddy, V. Durga Kumari, M. Subrahmanyam, Mater. Res. Bull. In press, (2009)

[5] J. Krishna Reddy, V. Durga Kumari, M. Subrahmanyam, presented at National Workshop on Catalysis, Bhubaneswar, 2008 (Unpublished)

[6] M. Noorjahan, V. Durga Kumari, M. Subrahmanyam, Appl. Catal. A. 234, 155 (2002)

[7] M. Noorjahan, V. Durga Kumari, M. Subrahmanyam, Appl. Catal. B. 47, 209 (2004)

[8] M. Noorjahan, V. Durga Kumari, M. Subrahmanyam, L. Panda, Appl. Catal. B 57, 291 (2005)

[9] M. Noorjahan, V. Durga Kumari, M. Subrahmanyam, J. Photochem. Photobiol. 156, 179 (2003)

[10] M. Noorjahan, V. Durga Kumari, M. Subrahmanyam, Ind. J. Environ. Protec. 22,1162 (2002)

[11] K. Lalitha, J. Krishna Reddy, M.V. Phanikrishna Sharma, V. Durga Kumari, M.Subrahmanyam, presented at National Workshop on Catalysis, Bhubaneshwar, 2008 (Unpublished)

[12] V. Durga Kumari, presented at National Seminar on ECOCHEM-2006, Nagpur, 2006 (Unpublished)

[13] K. Lalitha, J. Krishna Reddy, V. Durga Kumari, M. Subrahmanyam, presented at Catalysis for Sustainable Energy and Chemicals, CATSYMP-19, Pune, 2009 (Unpublished)

[14] A. Ratnamala, G. Suresh, V. Durga Kumari, M. Subrahmanyam, Mater. Chem. Phys. 110, 176 (2008)

[15] J. Krishna Reddy, G. Suresh, C. H. Hymavathi, V. Durga Kumari, M. Subrahmanyam, Catal.Today 141, 89 (2009)

[16] M.V. Phanikrishna Sharma, K. Lalitha, V. Durga Kumari, M. Subrahmanyam, Sol. Energy Mater. Sol. Cells 92, 332 (2008)

[17] M.V. Phanikrishna Sharma, V. Durga Kumari, M. Subrahmanyam, J. Haz. Mater. 160, 568 (2008)

[18] M.V. Phanikrishna Sharma, V. Durga Kumari, M. Subrahmanyam, Chemosphere 72, 644 (2008)

[19] M.V. Phanikrishna Sharma, A. Ratnamala, V. Durga Kumari, M. Subrahmanyam, PhD. Thesis, Osmania University, 2009

[20] M.V. Phanikrishna Sharma, V. Durga Kumari, M. Subrahmanyam, Chemosphere 73, 1562 (2008)

[21] M. Pratap Reddy, B. Srinivas, V. Durga Kumari, M. Subrahmanyam, P.N. Sharma, Toxicol. Environ. Chem. 86 (2004) 127.

[22] M. Pratap Reddy, N. Chandrasekar Rao, K. Krishna Prasad, K. Venkatamohan, P.N. Sharma, V. Durga Kumari, M. Subrahmanyam, Ind. J. Environ. Protec.22 (2002) 1253

[23] M. Subrahmanyam, P. Boule, V. Durga Kumari, D. Naveen Kumar, M. Sancelme, A. Rachel, Sol. Energy 82, 1099 (2008)

[24] M Pratap Reddy, H.H. Phil, M Subrahmanyam, Catal. Lett. 123, 56 (2008)

[25] M. Pratap Reddy, A. Venugopal, M. Subrahmanyam, Wat. Res. 41, 379 (2007)

[26] K.V. Subba Rao and M Subrahmanyam, Chem. Lett. 234 (2002)

[27] K.V. Subba Rao, B. Srinivas, A.R. Prasad, M. Subrahmanyam, Chem. Lett. 236 (2002)

[28] K.V. Subba Rao, B. Srinivas, A.R. Prasad, M. Subrahmanyam, JCS Che comm. 1533 (2000)

Mater. Res. Soc. Symp. Proc. Vol. 1171 © 2009 Materials Research Society   1171-S01-08

Effects of Synthesis Conditions on the Crystalline Phases and Photocatalytic Activities of Silver Vanadates Via Hydrothermal Method

Chao M. Huang[1], Guan T. Pan[2], Lung C. Chen[3], Thomas C.K. Yang[2], and Wen S. Chang[4]

[1] Department of Environmental Engineering, Kun Shan University, Yung Kang City, Taiwan

[2] Department of Chemical Engineering, National Taipei University of Technology, Taipei, Taiwan

[3] Department of Polymer Materials, Kun Shan University, Yung Kang City, Taiwan

[4] Energy and Environmental Laboratory, Industrial Technology Research Institute, Hsinchu, Taiwan

ABSTRACT

Visible-light-driven Ag_3VO_4 photocatalysts were successfully synthesized using low-temperature hydrothermal synthesis method. Under various hydrothermal conditions, the structures of silver vanadates were tuned by manipulating the hydrothermal time and the molar ration of silver to vanadium. X-ray diffraction (XRD) results reveal that the powders prepared in a stoichiometric ratio consisted of pure α-Ag_3VO_4 or mixed phases of $Ag_4V_2O_7$ and α-Ag_3VO_4. With increasing the Ag-to-V mole ratio to 6:1, the resulting samples were identified as pure monoclinic structure α-Ag_3VO_4. UV-vis spectroscopy indicated that silver vanadate particles had strong visible light absorption with associated band gaps in the range of 2.2-2.5 eV. The sample synthesized in the excess silver exhibited higher photocatalytic activity than that synthesized in a stoichiometric ratio. The powder synthesized at silver-rich at 140℃ for 4 h (SHT4) exhibited the highest photocatalytic activity among all samples. The reactivity of SHT4 (surface area, 3.52 $m^2 g^{-1}$) on the decomposition of gaseous benzene was about 16 times higher than that of P25 (surface area, 49.04 $m^2 g^{-1}$) under visible light irradiation. A well developed crystallinity of Ag_3VO_4 of SHT 4 was considered to enhance the photocatalytic efficiency.

INTRODUCTION

Silver vanadates, prepared by precipitation and solid-state reactions, have been investigated to be the photocatalysts for O_2 evolution [1]. Among the silver vanadates, only Ag_3VO_4 possessed the high visible-light photoactivity for O_2 evolution from an aqueous silver nitrate solution. With a strong oxidizing potential, Ag_3VO_4 is expected to be good candidate in the field of photodecomposition of organic compounds. However, the high temperature and long time reaction of solid-state reaction restricts further development in industrial production [2-4]. To reduce energy consumption, exploiting low cost metal oxide catalysts with high visible-light photocatalytic activity is desirable. Hence, the synthesis of silver vanadates using a one-step hydrothermal synthesis method is proposed in this study. The influences of the ratio of silver to vanadium as well as the hydrothermal times on the photocatalytic activities of Ag_3VO_4 were investigated.

EXPERIMENTAL DETAILS

In a typical experiment, first, 0.006 mol $AgNO_3$ was dissolved into 60 ml of de-ionized water under stirring; an appropriate NH_4VO_3 was dissolved in de-ionized water to form a transparent solution. Then, the aqueous $AgNO_3$ solution was added dropwise to NH_4VO_3 solution under vigorous stirring to make silver vanadium oxide suspension. The ratio of silver to vanadium in the starting solution was kept in a stoichiometric ratio (3:1) or in the presence of excess amounts of silver (6:1). The pH value of the whole solution was adjusted to 7 using ammonia solution and the solution was aged at room temperature for 24 h. Second, the suspended solution was poured into a 100 ml Teflon-lined stainless autoclave, sealed, heated to 140 ℃, and maintained for various hydrothermal times. After that, the autoclave was allowed to cool down naturally; the resulting precipitates were collected and washed several times with de-ionized water to remove NO_3^- and NH_4^+ residues, and finally dried at 80℃ for 6 h. The samples with the ratio of silver to vanadium in a stoichiometric ratio (3:1) were denoted as HT2, HT4, HT6, and HT8, indicating the hydrothermal times of 2, 4, 6, and 8 h, respectively. Besides, these samples in the presence of excess amounts of silver (6:1) were denoted as SHT2, SHT4, SHT6, and SHT8, respectively.

The *in-situ* DRIFT technique coupled with MS at the gas outlet was applied to simultaneously monitor the adsorption of VOCs on the catalyst surface and the composition of the effluent vapors. The spectra were recorded both in the dark and under illumination conditions. A fixed amount of VOC vapor was periodically withdrawn from the sampling port and injected into the MS to determine the gas composition. A schematic diagram of an in situ micro-reactor, a diffuse reflectance accessory used to reflect the time evolution of infrared spectrum, and the supply of VOCs, is shown in figure 1.

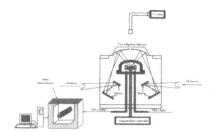

Figure 1. Schematic diagram of an in-situ micro-photoreactor.

DISCUSSION

Figure 2 shows the XRD patterns of silver vanadates prepared in a stoichiometric ratio. The HT2 and HT4 samples had mixed phases of $Ag_4V_2O_7$ and α-Ag_3VO_4. With increasing the hydrothermal time, XRD peaks for $Ag_4V_2O_7$ became weakened and XRD peaks for α-Ag_3VO_4 were gradually developed. Figure 3 shows the XRD patterns of silver vanadates prepared in the presence of excess silver. Apparently, all samples showed the XRD patterns indexed to α-Ag_3VO_4 (JCPDS 43-0542) and the diffraction peaks intensities are stronger than those of silver vanadates prepared in a stoichiometric ratio.

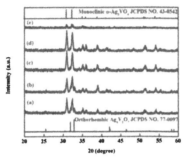

Figure 2. XRD patterns of samples for: (a) HT2, (b) HT4, (c) HT6, (d) HT8, and (e) using a precipitation method and calcined at 380 ℃ for 4 h in air.

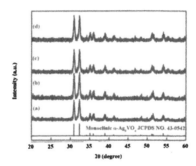

Figure 3. XRD patterns of samples for: (a) SHT2, (b) SHT4, (c) SHT6, and (d) SHT8.

The morphology of the as-prepared samples was investigated with SEM. Figure 4 showed that the morphology of the samples was strongly dependent on the ratio of silver to vanadium. Figure 4a shows the SEM of HT4 and HT4 samples had irregular ragged particles with sharp edges and less grain boundaries. In contrast, STH4 shows small particles with higher number of the grain boundaries (Fig. 4b). The specific surface areas of as-prepared powders were measured by nitrogen adsorption/desorption using the Barrett–Emmett–Teller (BET) method (Micromeritics ASAP 2020). SHT4 had specific surface area of 3.52 m^2 g^{-1}, whereas the HT4 had 2.04 m^2 g^{-1}.

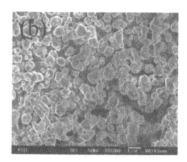

Figure 4. Morphologies of the samples: (a) HT4 and (b) SHT4.

The diffuse reflectance spectra of the as-prepared samples were measured with a JASCO V-500 spectrophotometer and converted by the Kubelka–Munk equation. The samples prepared in a stoichiometric ratio showed that the spectra patterns of HT6 and HT8 were similar (not shown). In addition, the values of band gap were estimated by the onset point of the absorption curve as 2.25-2.30 eV. However, the HT2 and HT4 samples had mixed phases of $Ag_4V_2O_7$ and α-Ag_3VO_4; the values of the band gap for HT2 and HT4 were found to be in the range of 2.40-2.50 eV.

The diffuse reflectance spectra of silver vanadates prepared in presence of excess silver were shown in figure 5. The spectra patterns of SHT6 and SHT8 were similar; the band gap absorption edge of SHT8 was determined to be 554 nm and the value of band gap estimated by the onset point of the absorption curve was 2.24 eV, which is similar to the value reported by Konta et al. [1].

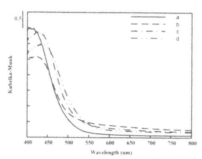

Figure 5. Diffuse reflectance spectrum of samples for: (a) SHT2, (b) SHT4, (c) SHT6, and (d) SHT8.

The photocatalytic activities of as-prepared samples were evaluated by the decomposition of gaseous benzene under visible-light irradiation. Each photoreaction was conducted in a continuous-flow annular photoreactor [5] with 0.05 g of sample evenly spread on a Pyrex tube. The light source with a wavelength ranging from 320 to 600 nm was used and the light intensity on the surface of powder was 0.2 mW/cm^2 with 3-5 % of the contribution of UV irradiation. The degradation data of the benzene were fitted to a pseudo first order rate model and the apparent rate constant k was obtained through linear fits of the equation: $\ln (C_e/C) = k_{app}t$ (where C_e and C are the equilibrium concentration of adsorption and the concentration of benzene at the irradiation time, t, respectively) [6]. The values of k_{app} of benzene degradation of as-prepared samples were shown in table I. It can be seen that the ratio of silver to vanadium has a significant effect on the photoactivity of the as-prepared samples. The samples prepared in silver-rich (Ag/V = 6:1) had the higher photocatalytic activities than those of samples prepared in a stoichiometric ratio. Moreover, the sample prepared in silver-rich (Ag/V = 6:1) at 140°C for 4 h (SHT4) exhibited the highest photocatalytic activity, 16.3 times higher than that of P25 for benzene, among the hydrothermal samples. Based on the results of XRD, Ag_3VO_4 prepared in excess silver had well crystallinity, resulting in an increase of photocatalytic activity.

Table I. Apparent rate constant of the samples.

Sample code	P25	HT2	HT4	HT6	HT8	SHT2	SHT4	SHT6	SHT8
k_{app} (min^{-1})	0.080	0.645	0.985	0.518	0.380	1.128	1.301	1.189	0.809

To investigate the adsorption/desorption characteristics, the *in situ* IR studies of the catalyst surface have been performed. Figure 6 shows the spectra obtained from the surface of HT4 samples. Figure 6a shows benzene adsorbed on the surface of HT4 prepared in a stoichiometric ratio during the 90 min of darkness. Due to measurement limitations, benzene adsorption on HT4 gave rise to v_{13} and v_{14} (C-H bending bands), while v_{13} (at 1484 cm^{-1}) and v_{14} (at 1038 cm^{-1}) bands reached a saturate state after 90 min [7]. Figure 6b shows the spectra of benzene adsorbed on the surface of SHT4, which had higher band intensities than those of HT4 in a stoichiometric ratio. The higher amount of adsorbed benzene may be due to the higher surface area.

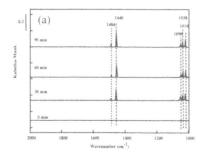 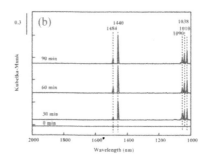

Figure 6. FTIR spectra of benzene adsorbed on (a) HT4 and (b) SHT4 samples during the 90 min of darkness.

The IR spectra of HT4 equilibrated with benzene vapors in the dark were taken as the initial state under illumination. The response was a result of photodecomposition and desorption of benzene. The DRIFT spectra of gaseous benzene recorded under visible-light irradiation for HT4 samples are shown in figure 7. Figure 7a indicated that the intensities of v_{13} and v_{14} decreased obviously after 10 min of illumination and two of those bands disappeared after 15 min of illumination. In contrast, the spectrum obtained at the same 10 min irradiation for SHT4 (figure 7b) is smaller than that of HT4 with stoichiometric ratio. Obviously, SHT4 had higher photocatalytic activity even though the SHT4 had much larger amount of benzene adsorbed on the surface at the initial state under illumination than that of HT4 (figure 7a).

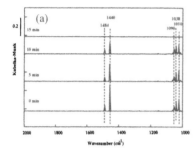 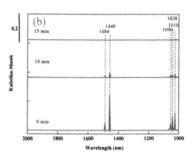

Figure 7. FTIR spectra of benzene adsorbed on (a) HT4 and (b) SHT4 samples after visible-light irradiation.

CONCLUSIONS

Due to the well developed crystallinity of Ag_3VO_4, the silver vanadates prepared in excess silver at 140℃ for 4 h showed the highest photocatalytic activity among all hydrothermal-synthesis samples. The reactivity of SHT4 (specific surface area, 3.52 $m^2 g^{-1}$) on benzene was 16.3 times higher than that of P25 (specific surface area, 49.04 $m^2 g^{-1}$) under visible light irradiation. The benzene adsorbed on SHT4 and concentration of gaseous benzene after visible-light irradiation for SHT4, detected by the *in-situ* FT-IR diffuse reflectance (DRIFT) technique, confirm the photocatalytic activity of SHT4.

REFERENCES

1. R. Konta, H. Kato, H. Kobayoshi, A. Kudo, *Phys. Chem. Chem. Phys.* **5**, 3061-3065 (2003).
2. J. Ye, Z. Zou, H. Arakawa, M. Oshikiri, M. Shimoda, A. Matsushita, T. Shishido, *J. Photochem. Photobiol. A: Chem.* **148**, 79-83 (2002).
3. X. Zhang, Z. Ai, F. Jia, L. Zhang, X. Fan, Z. Zou, *Mater. Chem. Phys.* **103**, 162-167 (2007).
4. H. Kato, H. Kobayashi, A. Kudo, *J. Phys. Chem. B* **106**, 12441-12447 (2002).
5. G.T. Pan, C.M. Huang, L.C. Chen, W.T. Shiu, *J. Environ. Eng. Manage.* **16(6)**, 413-420 (2006).
6. D. Gumy, S.A. Giraldo, J. Rengifo, C. Pulgarin, *Appl. Catal. B: Environ.* **78**, 19-29 (2008).
7. G.D. Lonardo, L. Fusina, G. Masciarelli, F. Tullini, *Spectrochim Acta A* **55**, 1535-1544 (1999).

Mater. Res. Soc. Symp. Proc. Vol. 1171 © 2009 Materials Research Society   1171-S02-01

Surface Modification of Tungsten Oxide-Based Photoanodes for Solar-Powered Hydrogen Production

N. Gaillard[1], J. Kaneshiro[1], E.L. Miller[1], L. Weinhardt[2,3], M. Bär[2,4], C. Heske[2], K. -S. Ahn[5], Y. Yan[5], and M. M. Al-Jassim[5]

[1]Hawaii Natural Energy Institute (HNEI), University of Hawaii at Manoa, Honolulu, HI 96822, USA
[2]Department of Chemistry, University of Nevada Las Vegas, NV 89154-4003, U.S.A.
[3]Experimentelle Physik II, Universität Würzburg, D-97074 Würzburg, Germany
[4]Helmholtz-Zentrum Berlin für Materialien und Energie, Lise-Meitner-Campus, D-14109 Berlin, Germany
[5]National Renewable Energy Laboratory (NREL), Golden, CO 80401, U.S.A.

ABSTRACT

We report on the development of tungsten-oxide-based photoelectrochemical (PEC) water-splitting electrodes using surface modification techniques. The effect of molybdenum incorporation into the WO_3 bulk or the surface region of the film is discussed. Our data indicate that Mo incorporation in the entire film (WO_3:Mo) results in poor PEC performances, most likely due to defects that trap photo-generated charge carriers. However, compared to a pure WO_3 (WO_3:Mo)-based PEC electrode, a 20% (100%) increase of the photocurrent density at 1.6 V vs. SCE is observed if the Mo incorporation is limited to the near-surface region of the WO_3 film. The resulting WO_3:Mo/WO_3 bilayer structure is formed by epitaxial growth of the WO_3:Mo top layer on the WO_3 bottom layer, which allows an optimization of the electronic structure induced by Mo incorporation while maintaining good crystallographic properties.

INTRODUCTION

More than three decades after the first report of photo-induced water splitting on TiO_2-based electrodes by Fujishima and Honda[1], intensive research is still ongoing to identify a suitable semiconductor to be integrated in an efficient, cost effective, and reliable photo-electro-chemical (PEC) system. Many candidate semiconductor materials have been studied for PEC applications, including amorphous silicon-based materials, III-V compounds, and transition metal oxides. One example of the latter is WO_3, which offers good corrosion resistance and is inexpensive to produce. However, the relatively high band gap of WO_3 (approx. 2.6 eV) results in a lack of sufficient absorption of the solar spectrum necessary for high hydrogen evolution rates. Several experiments have been performed at the University of Hawai'i to reduce the band gap of reactively-sputtered WO_3 films by incorporating impurities in the material bulk. For instance, a net 0.5 eV band gap reduction has been achieved using nitrogen as impurity[2]. However, photoelectrochemical characterizations performed in an 0.33M H_3PO_4 electrolyte under simulated AM1.5 global light showed a decrease of the saturated photocurrent density from ~2.68 mA.cm^{-2} (in WO_3) to ~0.67 mA.cm^{-2} (in WO_3:N). Additional SEM studies pointed out poor microstructural properties induced by nitrogen incorporation (using N_2 gas in addition to

O_2 during the reactive-sputtering process) that could affect both, bulk and surface electronic properties, most likely due to charge carrier trapping at defect sites organized in crystallographic shear (CS) planes[3]. One can see from this example that high performance WO_3-based PEC electrodes cannot be achieved by solely tailoring the band-gap, but that attention to the resulting bulk and surface electronic properties must also be paid. In this paper, we report our research focusing on the major component of a modified WO_3-based PEC electrode, studying both the absorber bulk and surface electronic properties. We present our most recent results on WO_3 band-gap tailoring with molybdenum (WO_3:Mo) using a co-sputtering process. In addition, promising results from bilayer-based PEC electrodes made of a thick WO_3 bottom film (absorber) and a relatively thin WO_3:Mo top layer (for band position shift) are discussed.

EXPERIMENTAL DETAILS

All samples studied in this paper were fabricated by reactive RF magnetron sputtering from metallic sputter targets (5 cm diameter) in a controlled argon/oxygen ambient. The substrates (SnO_2:F-coated glass) used in this experiment were cleaned in a soap/DI-water/acetone/DI-water/methanol/DI-water/isopropanol sequence. The substrates were mounted about 20 cm above the targets and heated to ~ 280°C. After reaching a base pressure in the 10^{-6} Torr range, argon and oxygen gas flow were set for a $[O_2]/([Ar]+[O_2])$ ratio of 25% at a fixed ambient pressure of 9.2 mTorr for both pure WO_3 and WO_3:Mo films. The films were synthesized using a W target powered at 250W, leading to a deposition rate in the 0.8-1.0 Å/s range. Molybdenum incorporation into WO_3 films was obtained by co-sputtering using a second Mo target powered at 200W, increasing the deposition rate to ~1.3 Å/s. This deposition process was used for all WO_3:Mo layers presented in this paper (thick layer in band-gap tailoring investigation and top part of the bilayer for band position shift). Before film deposition, all targets were pre-sputtered for 10 min with the shutters closed. The film thicknesses were measured by a stylus profilometer on several points of the samples. UV/vis transmission and reflection measurements were performed with a PE Lambda2 spectrophotometer with integrating sphere. The surface composition of the deposited films was characterized by x-ray photoelectron spectroscopy (XPS) using a Mg K_α x-ray source and a PHOIBOS 150MCD electron analyzer (SPECS). In addition, secondary ion mass spectroscopy (SIMS) analyses were performed on bilayer samples to evaluate the thickness of the WO_3:Mo top layer. UV photoelectron spectroscopy (UPS, using He I and He II excitation) and inverse photoemission spectroscopy (IPES, using a custom built setup with a high-flux low-energy electron gun (STAIB) and a Dose-type detector with a SrF_2 window and Ar:I_2 filling) were also performed to determine the valence and conduction band position at the surface, respectively (with respect to the Fermi energy). Finally, film morphologies were characterized using a Hitachi S-4800 scanning electron microscopy (SEM) for top view imaging and a Tecnai F20 UT high-resolution transmission electron microscope (HR-TEM) for cross-sectional analyses. For photoelectrochemical characterizations, photoelectrodes were prepared by soldering electrical leads with indium to the uncoated part of SnO_2:F bottom contacts on the substrate edges. All metallic parts (leads and indium) were then covered with epoxy to insulate them from the electrolyte. The electrode areas, typically ranging from 3 to 4 cm^2, were determined from optical scan images of the electrodes. The photoelectrodes were subjected to current-vs-potential scans in a three-electrode setup, using a 10 cm^2 Pt sheet counter electrode (CE) and a saturated calomel reference electrode (SCE), in

0.33M H_3PO_4 electrolyte (pH1.28). Current-vs-potential scans were recorded under simulated AM1.5 Global light produced by an Oriel 1-kW solar simulator.

RESULTS AND DISCUSSION

<u>1) Band-gap tailoring by Mo incorporation in WO_3</u>

The thicknesses of the here-studied WO_3 and WO_3:Mo samples were determined to be 2.2 µm and 2.1 µm, respectively. Surface XPS characterization using the O 1s, W 4p, and Mo 3d peak areas derived a [W]/([O]+[W]) ratio of ~23% and a [Mo]/([Mo]+[W]+[O]) ratio of ~7% for WO_3 and WO_3:Mo samples, respectively. Figure 1a presents the Tauc plots obtained from UV-vis transmission and reflection curves. In the case of WO_3, a 2.68 (±0.05) eV bulk band gap was obtained, a value close to others previously reported for this materials deposited with sputtering processes[4]. When molybdenum is incorporated in the WO_3 bulk, an identical band-gap (2.65 (±0.05) eV) is observed with optical techniques. The photoelectrochemical properties of WO_3 and WO_3:Mo-based PEC electrodes measured in a 0.33M H_3PO_4 solution under simulated chopped AM1.5G light are presented in Fig. 1b. For pure WO_3 films, a photocurrent density of 3.18 mA/cm^2 is achieved at 1.6 V vs. SCE, a value similar to those previously reported with this deposition process[4]. When molybdenum is incorporated in the entire WO_3 film (WO_3:Mo), the photocurrent density decreases to 1.91 mA/cm^2 (at 1.6 V vs. SCE). It is also interesting to note that the onset potential increases by 100 mV after Mo incorporation, from 475 mV vs. SCE for WO_3 to 575 mV vs. SCE for WO_3:Mo.

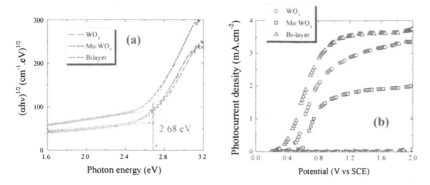

Fig. 1. (a) Tauc plots obtained from UV/vis absorption spectra measured on pure WO_3 (open red circles), WO_3:Mo (open blue squares), and WO_3:Mo/WO_3 bilayer (open green triangles). (b) J-V characteristics measured on WO_3 (open red circles), WO_3:Mo (open blue squares), and WO_3:Mo/WO_3 bi-layer (open green triangles) PEC electrodes in 0.33M H_3PO_4 under chopped simulated AM1.5G illumination (scan rate: 25 ms/V).

Accompanying SEM and TEM characterizations were performed to understand the decrease in performance after Mo incorporation. In the case of the WO_3 sample (Fig. 2a), the SEM micrograph shows a WO_3 polycrystalline film made of grains with sharp edges. When Mo is incorporated in the entire WO₃ film (Fig. 2b), the film surface is noticeably modified and exhibits smoother grains. From this analysis, one may assume that WO₃ crystallographic properties have been modified after Mo incorporation. In order to evaluate this hypothesis, TEM electron diffractions were performed on both WO₃ and WO₃:Mo films. To facilitate the comparison, both semi-diffraction patterns are presented together in Fig. 2d, with the WO₃:Mo semi-pattern displayed in the top part of the figure and that of WO₃ displayed in the bottom part. Note that the exact same area aperture (0.5 μm) was used for the diffraction pattern. It is clear from this characterization that Mo incorporation leads to a change in WO₃ film microstructure. In fact, the number of diffraction peaks measured on WO₃:Mo is less when compared to WO₃. Additional characterizations will be required to understand if this change in microstructure is related to the WO₃ monoclinic crystallographic phase.

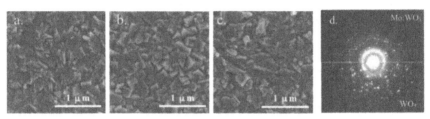

Fig. 2. SEM top view of (a) a WO₃, (b) a WO₃:Mo, and (c) a WO₃:Mo/WO₃ bilayer sample. Additionally, the montage of corresponding electron diffraction patterns obtained on WO₃Mo (top semi-pattern) and WO₃ (bottom semi-pattern) is shown in (d).

As was observed in the case of nitrogen incorporation into WO₃ films[2], molybdenum leads to a detrimental modification of both WO₃ morphological and crystallographic properties. The resulting WO₃:Mo-based PEC electrode shows poor performances, probably due to an increase in number of defects in a WO₃ material that already contains CS defect planes, well-known to limit electron transport properties[3].

2) Improving the PEC performance by Mo incorporation in the near-surface region of WO₃ films

WO₃:Mo/WO₃ bilayer samples were synthesized using the same processes used to fabricate WO₃ and WO₃:Mo films described in the previous section. To ensure that the top WO₃:Mo film of the bilayer has a similar stoichiometry as the WO₃:Mo thick film (i.e. [Mo]/([Mo]+[W]+[O] ratio of ~7%), the shutter was closed at the end of the WO₃ bottom layer deposition (while keeping the W target under bias), and then reopened after the 10 min pre-sputtering (i.e., closed-shutter) step of the Mo target (200W) to form the WO₃:Mo top layer. SIMS profiles measured on a bilayer sample indicated a top WO₃:Mo and a bottom WO₃ layers thickness of about 300 nm and 2 μm, respectively. The J-V characteristic of a WO₃:Mo/WO₃ bilayer-based PEC electrode measured in a 0.33M H₃PO₄ solution under simulated AM1.5 global chopped light is also presented in Fig. 1b. Clearly, the photoelectrochemical behavior changes when Mo is incorporated either into the entire bulk or only in the top part of the WO₃ film (bilayer). First, the onset potential of the bilayer-based PEC electrode is close to 360 mV vs. SCE, a 200 mV reduction compared to the WO₃:Mo material (575 mV vs. SCE). In addition, a photocurrent density of about 3.63 mA/cm² (at 1.6 V vs. SCE) is observed in the bilayer case, corresponding to an enhancement of 20% and 100% when compared to WO_3 and WO_3:Mo-based PEC electrodes (with comparable thickness), respectively. HR-TEM characterizations were performed to give additional insights for understanding the photoelectrochemical performance of the WO_3:Mo/WO_3 bilayer-based PEC electrodes. A cross section view of the interface is presented in Fig 3a. The position of this interface was previously estimated using a Z-contrast imaging TEM mode (note shown here). On this micrograph, both the WO_3 bottom and the WO_3:Mo top layer possess a high degree of crystallinity. Consequently, it is believed that the synthesis of an effective bilayer relies on the beneficial effect of the WO_3 bottom layer on the top WO_3:Mo film growth, as shown by the electron diffraction patterns (insets in Fig. 3a) indicating an epitaxial growth of the top WO_3:Mo layer. It is also worth mentioning that the resulting surface morphology of the WO_3:Mo/WO_3 bilayer samples observed by SEM and presented in Fig. 2c is very similar to the WO_3 one.

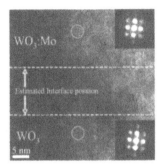 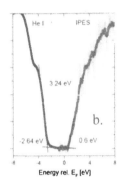

Fig.3. (a) HR-TEM micrograph performed at the interface between the bottom WO_3 (about 2µm thick) and the top WO_3:Mo (about 300 nm) layers. The insets show electron diffraction patterns acquired at the top and the bottom part (indicated by the circles) of the high-resolution image. (b) UPS and IPES spectra of a WO_3:Mo sample. For the IPES spectrum, the raw (grey line) and smoothed data (red dots) are shown. The linear extrapolation of edges gives the positions of the CBM and the VBM.

To conclude on the performances of the bilayer sample over the WO_3 one, it should be emphasized that comparable absorption characteristics have been experimentally observed in both the WO_3:Mo/WO_3 bilayer and the WO_3 films (Fig. 1a). Therefore, the charge collection must be higher in the WO_3:Mo/WO_3 bilayer electrodes compared to that of WO_3. This hypothesis is partially confirmed by UPS and IPES analyses performed on pure WO_3[5] and WO_3:Mo samples (Fig. 3b) indicating CBM and VBM upward shift after Mo incorporation. In fact, the CBM and VBM positions are found to be -2.89 eV vs. E_F[5] and -2.64 eV vs. E_F as well as +0.39 eV vs. E_F[5] and +0.6 eV vs. E_F for WO_3 and Mo:WO_3 films, respectively. Thus, when these two materials are in contact to form a bilayer, these energetic shifts would create a built-in electric field at the interface between the WO_3 bottom and WO_3:Mo top layers which would favor photo-generated holes transfer from the WO_3 bulk toward the electrode surface. Additional measurements of the electronic surface structure (also of actual WO_3:Mo/WO_3 bilayer samples) are necessary to confirm this hypothesis.

CONCLUSIONS

The results presented in this paper indicate that enhancement of tungsten-oxide-based PEC electrode performances cannot be obtained by reducing the band gap only. In fact, additional and potentially detrimental effects such as defect formation may arise form ion (N and Mo so far tested) incorporation into the WO_3 matrix. Therefore new concepts might be required to optimize WO_3-based material optical properties. In this study, it was demonstrated that better performance can be achieved by producing a bilayer structure of two different materials, selected for their absorption/transport and electronic properties, respectively. In the particular case of a bilayer formed by WO_3 and WO_3:Mo thin films, a 20% (50%) increase of the saturated photocurrent density was observed compared to pure WO_3 (WO_3:Mo)-based electrodes. High-resolution imaging techniques point out that a WO_3 film with a high degree of cristallinity can promote the growth of the subsequent WO_3:Mo top layer, collectively yielding an improved PEC performance.

ACKNOWLEDGMENTS

The work is supported by US Department of Energy under contract numbers DE-FC36-07GO17105 and DE-FG36-03GO13062. The authors would like to thank Brian Cole and Björn Marsen (past researchers at the University of Hawai'i) for their large contributions to the development of high performance WO_3-based PEC electrode as well as Xi Song, Alexander Deangelis, and Stewart Mallory from the University of Hawai'i for experimental assistance.

REFERENCES

[1] A. Fujishima, K. Honda, Nature 238, 37-38 (1972).
[2] B. Cole, B. Marsen, E. L. Miller, Y. Yan, B. To, K. Jones, and M. M. Al-Jassim, J. Phys. Chem. C 112, 5213-5220 (2008).
[3] Allpress, J. G.; Tilley, R. J. D.; Sienko, M. J. J. Solid State Chem. 1971, 3, 440.
[4] B. Marsen, E.L. Miller, D. Paluselli, R.E. Rocheleau, International Journal of Hydrogen Energy 32, 3110-3115 (2007).
[5] L. Weinhardt, M. Blum, M. Bär, C. Heske, B. Cole, B. Marsen, and E. L. Miller, J. Phys. Chem. C, 112, 3078-3082 (2008).

Mater. Res. Soc. Symp. Proc. Vol. 1171 © 2009 Materials Research Society   1171-S02-05

Effect of cationic or anionic dopants on optical and photocatalytic properties of TiO$_2$ nanopowders made by flame spray synthesis (FSS).

Katarzyna A. Michalow[1,2], Andre Heel[2], Thomas Graule[2], Mieczyslaw Rekas[1]
[1] Faculty of Material Science and Ceramics, AGH University of Science and Technology, al. Mickiewicza 30, 30-059 Krakow, Poland
[2] Laboratory for High Performance Ceramics, EMPA Swiss Federal Laboratories for Materials Testing and Research, Ueberlandstrasse 129, 8600 Duebendorf, Switzerland

ABSTRACT

TiO$_2$, TiO$_2$-1at.% W and TiO$_2$-1at.% Cr were produced from metal-organic precursors by flame spray synthesis (FSS). TiO$_2$-0.5at.% N was obtained by ammonolysis of FSS made TiO$_2$ nanopowder in a rotating tube furnace under NH$_3$ atmosphere. According to the X-ray diffraction (XRD) analysis, anatase is the predominant phase in all samples. Diffusive reflectance and the resulting band gap energy (E$_g$) were determined by diffusive reflection spectroscopy (DRS). Additional impurity bands at 2.43 and 2.57 eV for N- and Cr-doped TiO$_2$, respectively have been observed. The impurity band formed in the band gap resulted in increase of the light absorption in the visible range. The photocatalytic performance of the nanopowders under ultraviolet (UV, 290-410 nm) and visible light irradiation (Vis, 400-500 nm) was studied by the degradation of methylene blue (MB) in aqueous suspensions. It was found that all types of dopants influence the structure, interaction with the visible light as well as photocatalytic activity. Among all nanopowders, TiO$_2$-W exhibited the best photoactivity, much higher than the commercial TiO$_2$-P25 nanopowder. The optimum of the photodecolourization was obtained for 0.7 and 1 at.% W.

INTRODUCTION

Bare TiO$_2$ is the most common photocatalyst, due to its high corrosion resistivity and the non-toxicity. However, TiO$_2$ shows poor photocatalytic efficiency under solar light irradiation, due to its wide band gap and recombination losses of the photo charge pairs [1, 2]. There are several ways to overcome these drawbacks like doping of cationic sublattice of TiO$_2$ with either donor-type ions such as W^{6+}, Mo^{6+} and Nb^{5+} [3-5] or acceptor-type ions such as Cr^{3+}, Fe^{3+} [6, 7]. This type of modification by metal doping is particularly interesting to increase the electrical conductivity, resulting in a decrease of the recombination losses, as well as to modify the photocatalyst's surface properties leading to an improvement of the optical properties and relating electronic structure. Recently non-metal doping attracted considerable attention due to its enhanced photocatalytic activity in the visible light [8, 9]. An anionic doping by nitrogen seems to be the most interesting due to the noticeable red-shift of light absorption. New physical and chemical properties emerge when the size of the material becomes smaller and goes down to the nanometre scale. Well crystalline anatase particles with a size of about 11-21 nm are most suitable for a photocatalytic application [9-11].
The present study reports on the effect of: W^{6+}, Cr^{3+} and N^{3-} dopants on the optical and photocatalytic properties of TiO$_2$ nanoparticles.

EXPERIMENTAL DETAILS

Nanopowders of undoped and TiO_2 doped with W and Cr were produced by flame spray synthesis (FSS). Solution of titanium tetra-isopropoxide (TTIP; $Ti(C_3H_7O)_4$, 99%, VWR) in ethanol (EtOH, 99%, Sigma-Aldrich), tungsten hexacarbonyl (THC; $W(CO)_6$, 99%, ABCR) in tetrahydrofuran (THF; C_4H_8O 99.9% anhydrous, Aldrich) and chromium acetylacetonate (CHAA, $C_{15}H_{21}CrO_6$, 99%, ACROS) in m-xylene (1,3-Dimethylobenzene, C_8H_{10}, 99.9% anhydrous, ACROS) were used as a precursors of Ti, W and Cr, respectively. Details of synthesis and the setup of FSS are reported elsewhere [3, 4]. The precursors were mixed in proportions required to obtain desire composition and particle size of TiO_2 based materials.

The reaction of the oxides with ammonia gas (PanGas, 99.98 %, 2.5 $cm^3 \cdot s^{-1}$) was carried out at T = 550°C (heating rate: 10 $K \cdot min^{-1}$) in a rotating cavity quartz (SiO_2) reactor [12] with an internal diameter of 30 mm. The mass of sample of flame spray synthesised TiO_2 with a SSA = 54.1 $m^2 \cdot g^{-1}$ (referred to TiO_2-F54) was 0.25 g. During the reaction time of 4 hours, the ammonia was supplied through a quartz tube with a diameter of 5.8 mm placed above the sample. After the reaction the sample was quenched down to room temperature within 2 minutes under ammonia flow. The obtained sample was afterwards annealed at T = 350°C during 10 minutes under ambient air.

The primary particle size, shape and morphology of the particles were investigated by transmission electron microscopy (TEM; Philips CM30) operating at 300 kV.

X-ray diffraction analysis was performed with a Philips X'Pert Pro diffractometer using Cu Kα filtered radiation over a 2θ range from 20° to 80°. Phase identification was carried out using the database of the International Center for Diffraction Data (ICDD). The XRD results were used to determine the phase composition and resulting anatase/rutile fraction. The average crystallite size d_{XRD} and rutile-anatase ratio was determined.

The band gap energy E_g was determined with a Lambda 19 Perkin-Elmer double beam spectrophotometer. It is equipped with a 200 mm integrating sphere, used to measure the spectral dependence of total and diffuse reflectance over a wavelength λ of 250-2500 nm. Powder samples were measured in 1 mm thick quartz cell (Quartz Suprasil, Hellma). The reflectance of Spectralon (LabSphere) was used as a reference.

The photocatalytic activity (PCA) of the nanopowders was evaluated by the degradation of aqueous solution of methylene blue (MB) under UVA ($\lambda_{max} = 355$ nm) and Vis ($\lambda_{max} = 435$ nm) irradiation in the presence of the suspended photocatalyst. Setup of the photochemical reactor and experimental procedure has been reported elsewhere in detail [13, 14]. The decolorization rate of MB was evaluated by a Cary 50 UV-Vis spectrophotometer. Calibration of the absorbance at 664 nm against MB concentration was carried out for the determination of PCA and the concentration was determined from the peak intensity.

RESULTS AND DISCUSSION

The structure and morphology of flame made TiO_2-based nanopowders were investigated by several complementary methods and compared to TiO_2-P25 (Degussa), which is used as a reference. BET adsorption isotherm was applied to obtain the specific surface are (SSA) from which the mean grain particle (d_{BET}) has been calculated [4]. By means of X-ray diffraction (XRD) the structure and the phase composition was established as well as a mean crystallite size (d_{XRD}) [15]. XRD results revealed that both polymorphic forms of TiO_2, i.e., anatase (A) and

116

rutile (R), are present, but anatase dominates. No other phase than anatase and rutile were observed among all investigated samples. The varying amount of rutile fraction can be related to the fact that donor type of doping like W can act as an inhibitor of anatase to rutile (A-R) transformation [4, 16]. In contrast to this chromium can have the reversed effect i.e. it can promote A-R transformation, what was earlier reported by us for higher Cr concentrations [17]. In case of TiO$_2$-0.5 at.% N there is negligible increase of rutile fraction therefore we can assume that N-doping do not affect A-R transformation.

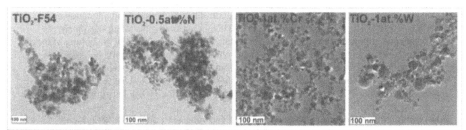

Figure 1. TEM images of undoped and doped TiO$_2$ nanopowders prepared by FSS.

TEM images (figure 1) from undoped, nitrogen, tungsten and chromium doped TiO$_2$ nanopowders show a typical particle morphology which can be expected from flame spray synthesis of metal-organic titanium precursors. The particles in the representative overviews are mainly spherical-shaped, non-aggregated and do not possess any interstitial particle necking, which is typical for such a kind of process. An afterwards ammonolysis has no effect on aggregation, whereas some particles rupture along the grain boundaries [13], what was confirmed by the decrease of the average crystallite size after ammonolysis. TEM pictures were also used to calculate mean grain size (d$_{TEM}$). Comparison of data from d$_{BET}$ and d$_{TEM}$ revealed a good agreement. Differences result from the measurement intrinsic assumptions, taken into account for the particle diameter calculations (table I) [18].

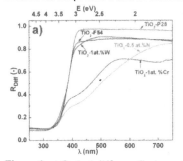 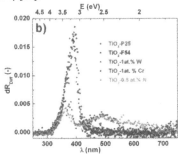

Figure 2. a) Optical diffuse reflectance R$_{Diff}$ as a function of wavelength λ and photon energy hν for TiO$_2$ and doped TiO$_2$ in comparison to TiO$_2$-P25, b) the first derivative spectrum dR$_{Diff}$/d(λ).

Figure 2a shows the optical diffuse reflectance spectra R$_{Diff}$ (λ) of undoped, Cr-, W- and N-doped TiO$_2$ nanopowders in comparison to TiO$_2$-P25. Pure TiO$_2$ both flame-made and commercial as well as W-doped TiO_2 exhibit a well-pronounced fundamental absorption edge that manifests itself as an abrupt drop in the R_{Diff} coefficient with the increasing wavelength. For TiO_2-F54 and TiO_2-1at.%W a decrease of R_{Diff} coefficient up from the fundamental absorption edge towards longer wavelengths. Doping by Cr and N results in an important deformation of TiO_2 spectra in the region close to the fundamental edge and a substantial decrease in the diffused reflectance. The band gap (E_g) calculation based on differential reflection spectra (figure 2b, table I). The photon energy at which the $dR_{Diff}/d(h\nu)$ attains its maximum corresponds to the energy of optical transitions defined as E_g [1]. Two optical transitions of energies (E_{gI}) 3.42-3.28 eV and (E_{gII}) 3.38-3.14 eV might be attributed to the band gap of anatase and rutile, respectively. The fundamental band gap of TiO_2 increases slightly when the dopant is introduced. The impurity band (E_{gd}) located at about 2.43 and 2.57 eV below the bottom of the conduction band of TiO_2 is related to nitrogen and chromium doping, respectively.

A summary of the most important results from our research of undoped, 1 at.%W, 1at.%Cr doped TiO_2 obtained by flame spray synthesis as well as TiO_2-0.5at.%N obtained by ammonolysis of TiO_2-F54 (FSS-made) nanopowder and reference TiO_2-P25 is given in table I.

Table I. Basic morphological parameters and band gap values of flame made TiO_2-based nanopowders in comparison to commercial nanopowder TiO_2-P25.

Sample	SSA $(m^2 \cdot g^{-1})$	d_{BET} (nm)	$d_{XRD,A}$ (nm)	$d_{XRD,R}$ (nm)	d_{TEM} (nm)	Rutile fraction (wt. %)	E_{gI} (eV)	E_{gII} (eV)	E_{gd} (eV)
TiO_2-P25	50*	-	20.58	35.99	-	16.9	3.35	3.14	-
TiO_2-F54	54.1	28.5	27.72	11.85	32.8	10.9	3.28	3.17	-
TiO_2-F106	106.8	16	13.4	6.7	-	9.1	3.31		-
TiO_2-0.5 at.% N	-	-	20.96	10.87	32.7	11.8	3.43	-	2.43
TiO_2-1 at.% Cr	87.1	17.8	13.8	9.3	18.8	9.0	3.42	3.38	2.57
TiO_2-1 at.% W	97.7	15.3	12	-	18.7	2.3	3.37	3.18	-

* Information by Degussa, A – anatase, R – rutile

The photocatalytic activity of flame made TiO_2-based nanopowders was evaluated on the basis of the photo-decolourisation rate of MB under UVA and Vis irradiation and compared to TiO_2-P25 nanopowder [19, 20]. The decomposition of MB under applied UV and Vis light in the range of 290-410 nm and 450-500 nm, respectively is assigned to photocatalytic activity of the investigated photocatalysts since the photosensitising effect can be neglected in this irradiation rage [19, 21].

The photodegradation kinetics were analyzed, by applying a Langmuir-Hinshelwood (LH) model and taking into account that the initial concentration C_0 (5.15×10^{-5} mol·dm^{-3}) was sufficiently low. Therefore, the simplified equation resulting from the L-H model can be written as:

$$\ln\left(\frac{C_0}{C}\right) = kKt = k_{app}t \qquad (1)$$

Where C is the concentration of the substrate (mol·dm^{-3}), t is the irradiation time (min), k is the rate constant (mol·dm^{-3}·min^{-1}), K the adsorption coefficient of the reactant (dm^3·mol^{-1}) and k_{app} is the first-order rate constant (min^{-1}) [22].

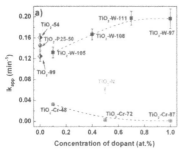
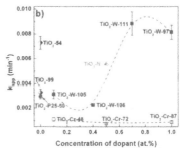

Figure 3. Comparison of the kinetic rate constant of decolourisation of MB **a)** for 15 min under UVA and **b)** for 180 min under Vis irradiation vs. the concentration of dopant in TiO_2. Used photocatalysts are marked in the following way: a letter indicates the dopant element; a number indicates SSA ($m^2 \cdot g^{-1}$).

As can be seen from figure 3, the photoreaction kinetic rate constant under both type of irradiation is the highest for the W-doped TiO_2 where with increasing W concentration the photoactivity (PCA) increase and the highest PCA can be observed for both 0.7 and 1 at.%W. The flame made TiO_2-F54 exhibited much higher photoactivity, in UV and Vis irradiation, if compared to TiO_2-P25 with the same SSA. In the case of the flame made TiO_2 an increase of the SSA has a negative impact on PCA. N-doping realized by ammonolysis of TiO_2-F54 decrease the photocatalytic performance in comparison to starting material. Increase of Cr concentration decreased PCA in both systems. The enhancement of the photoactivity for flame made TiO_2-F54 nanopowder is related to a higher crystallinity and larger crystallites size [10], what results in a lower number of defects. High number of defects may act as the electron traps but as well as the recombination centres, therefore the increase of the SSA resulted in decrease of PCA. On the other hand high PCA of flame made TiO_2 is caused by higher amount of anatase phase in comparison to TiO_2-P25. Anatase is known to be the favourable structure for photocatalytic applications due to the suitable position of its conductive band edge regarding to the reduction potential of water as well as a better affinity of the anatase surface to water and hydroxyl group adsorption [11]. Better performance of flame made TiO_2 and TiO_2-W can be also related to the increase of the light absorption, expressed by the increase of R_{Diff} coefficient, while the fundamental absorption edge is not affected at all. Cr as well as N doping had a significant influence on the fundamental absorption edge (figure 2), resulting from additionally introduced bands within the band gap. The impurity bands may act as easier paths for the excited electrons which in theory should increase efficiency of the light absorption. On the other hand additional bands may act as recombination centres and this effect can be observed in case of Cr and N doping.

CONCLUSIONS

Effect of cationic and anionic type of doping on the optical properties and photoactivity of TiO_2 nanopowder has been investigated. It becomes obvious, that there is a correlation reversal between the band gap modification and photodecomposition of methylene blue (MB) under UVA and Vis irradiation. Cr- and N- doping significantly affected the fundamental band edge by introducing impurity bands, where W-doping decreased the optical diffuse reflectance.

Flame made TiO_2 with SSA = $54m^2 \cdot g^{-1}$ and 0.7 and 1at.%W nanopowders exhibited the highest photoactivity 1.1 and 2.4 times higer in case of undoped TiO_2 and 1.35 and 3 times higher in case of W doped powders that TiO_2-P25 under UV and Vis irradiation, respectively. Cr and N doping as well as increase of the SSA for flame made TiO_2 showed a negative impact on the photocatalytic kinetic rates.

ACKNOWLEDGMENTS

The authors would like to acknowledge InnoGrant (the supporting program of innovative activity of PhD students) and the Polish Ministry of Science and Higher Education (N N508 1845 33) for the financial support.

REFERENCES

1. M. Radecka, M. Rekas and K. Zakrzewska, *Trends Inorg. Chem.* **9,** 81 (2006).
2. A. Fujishima and X. Zhang, *C. R. Chim.* **9**, 750 (2006).
3. K. Akurati, A. Vital, J.-P. Dellemann, K. Michalow, T. Graule, D. Ferri and A. Baiker, *Appl. Catal., B* **79**, 53 (2008).
4. K.A. Michalow, A. Vital, A. Heel, T. Graule, F.A. Reifler, A. Ritter, K. Zakrzewska and M. Rekas, *J. Adv. Oxid. Technol.* **11**, 56 (2008).
5. X.Z. Li, F.B. Li, C.L. Yang and W.K. Ge, *J. Photochem. Photobiol., A* **141**, 209 (2001).
6. E. Borgarello, J. Kiwi, M. Graetzel, E. Pelizzetti and M. Visca, *J. Am. Chem. Soc.* **104**, 2996 (1982).
7. M. Radecka, K. Zakrzewska, M. Wierzbicka, A. Gorzkowska and S. Komornicki, *Solid State Ionics*, **157**, 379 (2003).
8. R. Asahi, T. Morikawa, T. Ohwaki, K. Aoki and Y. Taga, *Science*, **293**, 269 (2001).
9. X. Chen and S.S. Mao, *Chem. Rev.* **107**, 2891 (2007).
10. C.B. Almquist and P. Biswas, *J. Catal.* **212**, 145 (2002).
11. A. Sclafani and J.M. Herrmann, *J. Phys. Chem.* **100**, 13655 (1996).
12. D. Logvinovich, A. Borger, M. Dobeli, S.G. Ebbinghaus, A. Reller and A. Weidenkaff, *Prog. Solid Chem.* **35,** 281 (2007).
13. K.A. Michalow, D. Logvinovich, A. Weidenkaff, M. Amberg, G. Fortunato, A. Heel, T. Graule and M. Rekas, *Catal. Today*, **doi:10.1016/j.cattod.2008.12.015** (2009).
14. A. Mills, S. Morris and R. Davies, *J. Photochem. Photobiol., A* **70**, 183 (1993).
15. K.K. Akurati, A. Vital, U.E. Klotz, B. Bommer, T. Graule and M. Winterer, *Powder Technol.* **165**, 73(2006).
16. S. Komornicki, M. Radecka and P. Sobas, *Mater. Res. Bull.* **39**, 2007 (2004).
17. A. Trenczek-Zajac, M. Radecka, M. Jasinski, K.A. Michalow, M. Rekas, E. Kusior, K. Zakrzewska, A. Heel, T. Graule and K. Kowalski, *J. Power Sources*, **doi:10.1016/j.jpowsour.2009.02.068**
18. M. Radecka, M. Rekas, E. Kusior, K. Zakrzewska, A. Heel, K.A. Michalow and T. Graule, *J. Nanosci. Nanotechnol.* **accepted for printing** (2009).
19. X. Yan, T. Ohno, K. Nishijima, R. Abe and B. Ohtani, Chemical Physics Letters, 429 (2006) 606.
20. H. Lachheb, E. Puzenat, A. Houas, M. Ksibi, E. Elaloui, C. Guillard and J.-M. Herrmann, *Appl. Catal., B* **39**, 75 (2002).
21. A. Mills and J. Wang, *J. Photochem. Photobiol., A* **127**, 123 (1999).
22. M.A. Fox and M.T. Dulay, *Chem. Rev.* **93**, 341 (1993).

Mater. Res. Soc. Symp. Proc. Vol. 1171 © 2009 Materials Research Society   1171-S03-05

Development of a hybrid photoelectrochemical (PEC) device with amorphous silicon carbide as the photoelectrode for water splitting

Jian Hu[1], Feng Zhu[1], Ilvydas Matulionis[1], Todd Deutsch[2], Nicolas Gaillard[3], Eric Miller[3], and Arun Madan[1]

[1]MVSystems, Inc., 500 Corporate Circle, Suite L, Golden, CO, 80401, USA
[2]National Renewable Energy Laboratory (NREL), Golden, CO 80401, USA
[3]Hawaii Natural Energy Institute (HNEI), University of Hawaii at Manoa, Honolulu, HI 96822, USA

ABSTRACT

We report on an integrated photoelectrochemical (PEC) device for hydrogen production using amorphous silicon carbide (a-SiC:H) material as the photoelectrode in conjunction with an amorphous silicon (a-Si:H) tandem photovoltaic device. With the use of a-Si:H tandem solar cell, the flat-band potential of the hybrid PEC structure shifts significantly below the H_2O/O_2 half-reaction potential and is in an appropriate position to facilitate water splitting. Under reverse bias, saturated photocurrent of the hybrid device ranges between 3 to 5 mA/cm^2 under AM1.5 light intensity. In a two-electrode setup (with ruthenium oxide counter electrode), which is analogous to a real PEC configuration, the hybrid cell produces photocurrent of about 0.83 mA/cm^2 at zero bias and hydrogen production is observed. The hybrid device exhibits good durability in pH2 buffered electrolyte for up to 150 hours (so far tested).

INTRODUCTION

Hydrogen is emerging as an alternative energy carrier to fossil fuels due to its non-toxic and environmentally friendly nature. Compared with other conventional methods (e.g., direct electrolysis, thermo-chemical decomposition of water and so on), photoelectrochemical (PEC) water splitting at a semiconductor-electrolyte interface using sunlight is of considerable interest as it offers an environmentally "green" approach to hydrogen production [1.2]. An efficient PEC water splitting device requires the semiconductor material used as a photoelectrode fulfilling a number of primary requirements, such as bandgap (E_g), band edge alignment and corrosion resistance to electrolyte. We have previously reported on a PEC device with amorphous silicon tandem junction with a WO_3 photoanode, leading to a 3% solar-to-hydrogen (STH) efficiency [3]; we have also reported a similar PEC device with the hydrogenated amorphous silicon carbide (a-SiC:H) photoelectrode fabricated by the plasma enhanced chemical vapor deposition (PECVD) technique [4,5]. Compared with conventional large bandgap photoelectrode materials (e.g., WO_3 with E_g of ~2.8eV), a-SiC:H film with a lower E_g (2.0-2.3 eV) would absorb more photons from sunlight and thus should enhance the STH efficiency. It should be noted that incorporation of the C in the film should lead to an increase in the corrosion resistance compared to the use of conventional a-Si:H films, which have poor stability in the electrolyte [6]. Moreover, a-SiC:H films are routinely prepared by the PECVD technique which is normally used in the fabrication of a-Si tandem solar cells. This could become very attractive for mass-production of PEC devices in a cost-effective fashion.

In this paper, we focus on the study of the hybrid PEC device consisting of the a-Si:H tandem solar and the a-SiC:H photoelectrode which has exhibited H_2 generation and excellent durability in an acidic electrolyte. We will also discuss the roadmap to achieve STH >10%, considered the minimum efficiency for economically viable hydrogen production, using an integrated PV/a-SiC:H hybrid device.

EXPERIMENTAL DETAILS

The intrinsic a-SiC:H films were deposited using H_2, CH_4 and SiH_4 gas mixtures at 200°C, in a PECVD cluster tool system specifically designed for the thin film semiconductor market and manufactured by MVSystems, Inc. The flow rates were varied from 0 to 12 sccm for CH_4. During the deposition, the pressure used was in a range of 400 to 900 mTorr and the RF power density used was in the range of 8-20 mW/cm^2. For both a-Si:H and a-SiC:H p-i-n solar cells, the p-layer was a-SiC:H:B and was deposited from SiH_4, CH_4, and B_2H_6 gas mixtures while the n-layer was prepared using SiH_4 and PH_3 gas mixture. The optoelectronic properties of a-SiC:H films such as dark and photoconductivity, and the solar cell performance such as the short-circuit current (J_{sc}), open-circuit voltage (V_{oc}), fill-factor (FF) and efficiency were characterized using a calibrated Global AM 1.5 light source (Xenon lamp). For the spectral response (quantum efficiency) measurement, narrow bandwidth filters were used. The Si-C bonding configurations in a-SiC:H films were determined by infrared (IR) spectroscopy .

The configuration used for the hybrid PEC device is shown in Fig.1 and consists of an a-Si:H tandem solar cell (thickness of the top and bottom cell is 80 nm and 360 nm respectively)

Fig.1. Configuration of the hybrid PEC device.

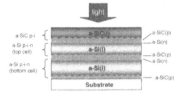

and a-SiC:H photoelectrode which contains an intrinsic a-SiC:H (~100 nm thick) and a thin p-type a-SiC:H:B layer (~10 nm thick). The substrate used for the hybrid PEC device was typically fluorine doped tin oxide (Asahi U-type) coated glass. Other types of substrates such as stainless steel (SS) and zinc oxide (ZnO) coated glass were also used for comparison purposes. In general, the a-SiC:H photoelectrode behaves like a photocathode where the photogenerated electrons inject into the electrolyte at the a-SiC(i)/electrolyte interface to reduce H^+ ions for hydrogen evolution. This way, anodic reactions and thus corrosion on a-SiC:H layer can be mitigated. The a-Si:H tandem solar cell which was used in the hybrid PEC device exhibited V_{oc} = 1.66 V, J_{sc}=8.7 mA/ cm^2, FF=0.67, and efficiency of ~9.6%.

The PEC performance of the hybrid device was evaluated at the National Renewable Energy Laboratory (NREL) and the Hawaii Natural Energy Institute (HNEI).

RESULTS AND DISCUSSION

A-SiC:H material and photoelectrode

Fig.2(a) shows the bandgap, E_g, of a-SiC:H materials as a function of the gas ratio $CH_4/(CH_4+SiH_4)$ used during the deposition. Note E_g was deduced using Tauc's plot [7] from measured transmission and reflectance data since the optically induced transition in a-SiC:H is a direct transition. As this gas ratio increases from 0.2 to ~0.5, E_g increases nearly linearly from 1.98 eV to ~2.2 eV, due to increasing C incorporation. Typically, for a-SiC:H films with E_g=2 eV, the C concentration as determined by X-ray photoelectron spectroscopy (XPS) is ~7%. It should be noted that as E_g increases to ~2.2 eV, the density of defect states (DOS) remains low as is indicated by the γ factor which remains in the range of 0.9 to 1. (Here, the γ factor is defined from $\sigma_{ph} \propto F^\gamma$, where σ_{ph} is the photoconductivity and F is the illumination intensity; we infer the DOS of the amorphous semiconductor from this measurement [8]). The good quality of the a-SiC:H was also confirmed by incorporating the material into a normal solar cell in the configuration, glass/Asahi U-Type SnO₂/p-a-SiC:B:H/i-a-SiC:H/n-a-Si/Ag. Such a device (0.25 cm²) exhibited a conversion efficiency, η, of 6.9% (V_{oc} = 0.91 V, J_{sc}=11.64 mA/ cm², FF=0.66) [4,5] and is close to the previously reported a-SiC:H solar cell [9].

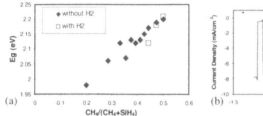 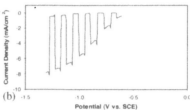

(a)

(b)

Fig.2. (a) E_g vs. CH₄/(CH₄+SiH₄); (b) the current density vs. potential characteristic of the a-SiC;H photoelectrode with the 200-nm a-SiC(i) layer (measured in 0.5M H₂SO₄ electrolyte).

The a-SiC:H photoelectrode is of a-SiC(p+)/a-SiC(i) configuration. The thickness of the a-SiC(p+) and a-SiC(i) layers is typically ~20 nm and 100-200 nm, respectively. The current density vs. potential characteristic (vs. SCE) of a 200-nm a-SiC:H photoelectrode is shown in Fig.2(b). It is seen that the a-SiC:H photoelectrode approaches a saturated photocurrent of ~8 mA/cm² (@-1.3V vs. SCE). Good durability of the a-SiC:H photoelectrode for up to 100 hours in pH2 electrolyte has been demonstrated [5].

Flat-band potential

Fig.3 shows the flat-band potential (V_{fb}), determined by the illuminated open-circuit potential (OCP) method, as a function of pH of electrolyte for both the a-SiC:H photoelectrode (open circles) and the hybrid PEC device (open triangles). The change of the V_{fb} with pH nearly exhibits a slope of ~60 mV/pH, as predicted by the Nernst equation [10]. We note in Fig 3, that at pH=2, V_{fb} = +0.26 V (vs. Ag/AgCl) for the a-SiC:H photoelectrode, whereas in the case of the hybrid device, the V_{fb} significantly by ~ +1.6 V (as indicated by the red

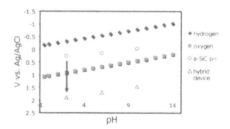

Fig.3. V_{fb} vs. pH for the a-SiC:H photo-electrode and the hybrid PEC device.

123

arrow) and is below the H_2O/O_2 half-reaction potential (by +0.97 V) and is in an appropriate position to facilitate water splitting.

Photocurrent of the hybrid PEC device

Figs.4(a) and (b) show the current density vs. potential characteristics for hybrid PEC device fabricated on different substrates, SnO_2 and ZnO coated glass and SS, and measured in the pH2 buffered electrolyte (sulphamic acid solution with added potassium biphthalate) using the 3-electrode and 2-electrode setup respectively. In the 2-electrode setup, there was no reference electrode and contained only the working electrode (hybrid PEC device) and the counter electrode which was ruthenium oxide (RuO_2) rather than conventional platinum (Pt).

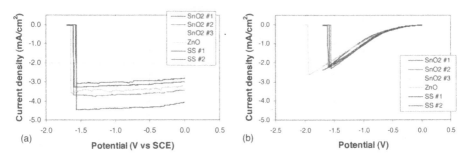

(a)

(b)

Fig.4. Current density vs. potential characteristics measured in (a) 3-electrode and (b) 2-electrode setup.

From Fig.4(a), we see the saturated photocurrent of the hybrid cell using different substrates is in the range of 3-5 mA/cm^2. The larger photocurrent using SnO_2 (>4 mA/cm^2) coated glass substrate is due to the inherent texture of the SnO_2 which enhances internal photon absorption. More significantly, we see that the photocurrent density of ~0.3 mA/cm^2 occurs at a zero potential using the 2-electrode setup. Hydrogen production was observed in a short-circuit condition, as shown in Fig.5.

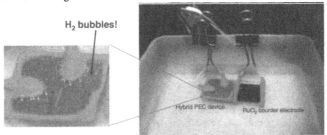

Fig.5. Hydrogen evolution occurs in the hybrid PEC device operated in the short-circuit condition.

It should be noted that compared with the 3-electrode case (Fig.4(a)), the photocurrent measured in the 2-electrode setup (even using RuO_2 counter electrode) is much lower, suggesting limiting factors. We have noted that the over-potential loss in the 2-electrode setup can be due to, (1) type of electrolyte used, (2) type of counter electrode used and (3) formation of thin $SiOx$ layer on the a-SiC surface. Initial results have shown that, after dipping the hybrid device into 5% hydrofluoric (HF) acid for 30 seconds and using RuO_2 as the counter electrode, the photocurrent is enhanced from 0.33 to 0.83 mA/cm^2 at zero bias.

Durability of the hybrid PEC device

The test was performed in the pH2 buffered electrolyte, with Pt as the counter electrode. During test, a constant current density of 1.6 mA/cm^2 was applied to the device, while the voltage (potential) across the sample was recorded over a 148-hr period. The current density vs. potential characteristics of the device prior to and after 22, 48 and 148-hr tests were measured in both the 3- and 2-electrode setups. Throughout these durability tests, H_2 production from the hybrid device occurred. Fig.6(a) shows the current vs. potential curves measured prior to and

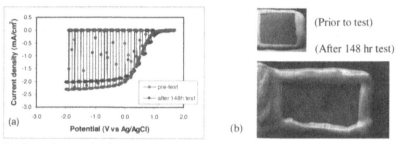

Fig.6. (a) Current density vs. potential characteristics measured prior to and 148 hr test. (b) Photo images of the hybrid PEC device prior to and after a 148-hr test in pH2 electrolyte.

after 148-hr test. These results show that the dark current shows almost no change, and hence no corrosion occurs in the hybrid device, after the 148-hr test, as is evident also in Fig.6(b).

Solar–to-hydrogen efficiency >10% -a roadmap

As determined by the U.S. Department of Energy, the technical objective defined for the PEC Hydrogen Production, using Silicon-based photoelectrodes, requires a STH, efficiency ~5% for the near-term (by 2009) and 10% by 2018 [11]. For Global AM1.5 solar irradiance, the STH efficiency can be defined as,

$$\text{Efficiency} = \frac{\text{chemical energy in hydrogen produced in a PEC device}}{\text{energy in the sunlight over the collection area}} \approx 1.23 \times J_{ph}, \qquad (1)$$

where J_{ph} is the photocurrent produced per unit irradiance area in a PEC device. From Fig.4(a), we have shown that the hybrid PEC device using the existing configuration leads to photocurrent

~4.5 mA/cm^2, potentially leading to a STH efficiency of ~5.5%. Further enhancement in the STH efficiency could be achieved by employing a nano-crystalline Si p-i-n solar cell (used as the bottom cell) in the tandem solar cell. We expect that with such a configuration, a photocurrent of ~8 mA/cm^2 could be generated, leading to a STH efficiency up to ~10%.

CONCLUSIONS

Hydrogen production has been demonstrated in the hybrid PEC device consisting of the a-Si:H tandem solar and the a-SiC:H photoelectrode. The hybrid PEC exhibits good durability in acidic electrolyte for up to ~150 hours (so far tested). It is found that the SiOx layer on the surface of the a-SiC:H electrode, type of electrolyte and the counter-electrode contribute to the over-potential losses and thus hinder the photocurrent. Finally, we show that it may be possible to enhance the solar to hydrogen (STH) efficiency >10%.

ACKNOWLEDGMENTS

The work is supported by US Department of Energy under contract number DE-FC36-07GO17105. The authors would like to thank Ed Valentich for his assistance in sample fabrication, J. Gallon for his assistance in conductivity activation energy measurement and Dr. A. Kunrath for helpful discussions.

REFERENCES

1. T. Ohta, "Solar-Hydrogen Energy Systems", (Pergamon Press, 1979).
2. R. Narayanan and B. Viswanathan, "Chemical and Electrochemical Energy Systems", (University Press, India, 1998).
3. A. Stavrides, A. Kunrath, J. Hu, R. Treglio, A. Feldman, B. Marsen, B. Cole, E.L. Miller, and A. Madan, Proc. SPIE, Vol. 6340, 63400K(2006).
4. F. Zhu, J. Hu, A. Kunrath, I. Matulionis, B. Marsen, B. Cole, E.L. Miller, and A. Madan, Proc. SPIE, Vol. 6650, 66500S(2007).
5. J. Hu, F. Zhu, I. Matulionis, A. Kunrath, T. Deutsch, L. Kuritzky, E. Miller, and A. Madan, Proc. 23rd European Photovoltaic Solar Energy Conference, 69(2008)
6. K. Varner, S. Warren and J. A. Turner, Proceedings of the 2002 U.S. DOE Hydrogen Program Review, NREL/CP-610-32405.
7. J. Tauc, "Amorphous and liquid semiconductor" (Plenum, New York,1974) p.159
8 A. Madan and M.P. Shaw, "The Physics and Applications of Amorphous Semiconductors", (Academic Press, 1988) p. 99.
9. R.E.Hollingsworth, P.K.Bhat, and A.Madan, Proc. 19th IEEE PVSC, 684(1987).
10. R. Memming, "Semiconductor Electrochemistry", (Wiley-VCH, 2001).
11. Statement of Project Objectives, US Department of Energy, #DE-FC36-07GO17105; See also R. Garland et al, DOE Hydrogen Production Review Meeting, Arlington, VA, June 12, 2008.

Mater. Res. Soc. Symp. Proc. Vol. 1171 © 2009 Materials Research Society   1171-S04-01-Q05-01

Polymer-Titania Composites for Photocatalysis of Organics in Aqueous Environments

Cecil A. Coutinho and Vinay K. Gupta
Department of Chemical & Biomedical Engineering
University of South Florida, Tampa, FL – 33620

ABSTRACT

Microcomposites composed of titanium dioxide nanoparticles embedded within cross-linked, thermally responsive microgels of poly(N-isopropylacrylamide) were prepared. These microcomposites showed rapid sedimentation, which is useful for gravity separation of the titania nanoparticles in applications such as environmental remediation. To investigate the degradation kinetics using these microcomposites in aqueous suspensions, methyl orange was employed as a model contaminant. The decline in the methyl orange concentration was monitored using UV-Vis spectroscopy. Degradation of methyl orange was also measured using only nanoparticles TiO_2 (Degussa[TM] P25) for comparison with the microcomposites. Experiments were performed at different pH conditions that spanned acidic, neutral, and basic conditions to gain insight into the interplay of TiO_2 surface charge, ionization of the polyelctrolyte chains in the microcomposites, and ionization of the methyl orange.

INTRODUCTION

Titanium dioxide is a common and widely studied photocatalyst due to its appealing attributes such as non-toxicity, chemical inertness and high photocatalytic activity[1-3]. The large band gap of TiO_2 (3.2eV) permits it to absorb photons in the UV region, which results in production of electron-hole pairs that participate in redox reactions known to degrade simple organic species[4]. In recent years, there has been increased interest in use of nanosized titania powders due to enhancements in photocatalytic activity[5-7]. Because separation of suspended titania nanoparticles from water has been a major obstacle, use of very fine particles of titania in applications such as waste-water treatment have been limited. In a recent report[8], we demonstrated the synthesis of novel microcomposites in aqueous media comprising of polymer gels on micron length scales that were loaded with Degussa[TM] P25 TiO_2 nanoparticles. Microcomposites with high mass fractions of titania (50-75wt%) were prepared that showed rapid sedimentation (~minutes), which is a useful characteristic for gravity separations. In this study we build on our previous body of work by investigating the photodegradation of a model organic dye, methyl orange, using the novel polymer-titania microcomposites. Kinetics of the photodegradation are evaluated by monitoring the changes in the methyl orange concentration using UV-Vis spectroscopy. The influence of pH of the solution, which influences the interactions between the poly(acrylic acid) (PAAc) in the polymer microgels, the titania surface and the methyl orange adsorbate is studied. Degradation of methyl orange using freely suspended titania is also conducted for comparison with the microcomposites.

EXPERIMENTAL: PHOTOCATALYTIC STUDY

Degradation studies were done using aqueous solutions containing 5 ppm of methyl orange (MO) and suspended microcomposites or free Degussa[TM] P25 particles such that the overall concentration of TiO_2 was 50, 100, or 200 ppm. The pH was adjusted using 0.1M HCl or 0.1M NaOH and degradation kinetics are reported here at a pH of 2±0.1 and 6.5±0.2. Photocatalysis was performed under illumination using two commercially available 15W Philips F15T8 black-light fluorescent bulbs (model 392233) that have spectral energy distribution centered at 352 nm (UVB radiation). The intensity of the radiation reaching the solution surface ($3.5mJ/cm^2$) was detected using a Chromaline UV Minder radiometer (UVM226) connected to a remote probe (UVM226S). Control experiments were performed under UV irradiation without the addition of any catalyst or microcomposites. Negligible decolorization (<1%) of the MO was observed confirming that the degradation of MO predominately occurs by photocatalysis using titania rather than photolysis. Dark adsorption was conducted for at least three hours before irradiation for adsorption of MO onto the TiO_2 surface. Aliquots of 1.5ml of the suspension were collected at regular intervals during the degradation experiments, centrifuged (10000g, 30mins) to completely remove any particles, and the peak absorbance was analyzed using a V-530 UV-Vis spectrophotometer (Jasco, MD).

RESULTS AND DISCUSSION

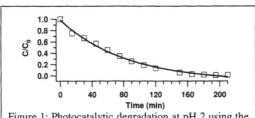

Figure 1: Photocatalytic degradation at pH 2 using the microcomposites (square). Line represents 1st order fit.

The microcomposites used in this study consist of porous, cross-linked PNIPAM microspheres with interpenetrating chains of PAAc that allow easy and efficient loading of titania nanoparticles within the IP-microgels[8]. The degradation of MO by both the microcomposite particles and freely suspended titania (Figure 1) is well described by a mono-exponential curve, suggesting that a pseudo-first-order reaction model can be used for describing the kinetic behavior of the photocatalysis. The apparent rate constant was obtained directly using a regression analysis of the experimentally observed decline in the peak height of MO absorbance as a function of time.

Table 1: Degradation rate constants

Concentration (PPM)	Freely Suspended TiO_2		Microcomposite	
	$k\ (min^{-1})\ pH\ 2$	$k\ (min^{-1})\ pH\ 6.5$	$k\ (min^{-1})\ pH\ 2$	$k\ (min^{-1})\ pH\ 6.5$
200	0.0130	0.0189	0.0136	0.0046
100	0.0093	0.0154	0.0084	0.0053
50	0.0036	0.0086	0.0033	0.0036

Table 1 shows that the reaction rate constant for the photodegradation of MO by the freely suspended TiO_2 at the acidic pH (~2) is 57% higher than the value obtained when the pH is near neutral conditions (~6.5). This effect has been reported in literature[9, 10] and can be attributed to the increased active sites available for MO degradation. Regarding the pH variation our results correlate well with that of Kansal and co-workers[11], who also reported faster degradation kinetics under neutral (or basic) conditions using freely suspended TiO_2. At a basic pH, both the surface of the titania (ISP ~ 6.5) and the

MO are negatively charged, which impacts surface adsorption of MO. However, the presence of large quantities of hydroxyl ions on the particle surface as well as in the reaction medium favors the formation of OH• radicals that are responsible for the photocatalytic degradation of MO. Even though the concentration of OH- ions in bulk solution is reduced at a neutral pH, the electrostatic repulsion between the titania surface, which is now relatively uncharged, and the MO is reduced favoring adsorption of the dye and its oxidation. In an acidic medium, although the adsorption of the MO is improved on the positively charged TiO_2 surface, the reduction in the concentration of the OH- ions leads to a decrease in the rate constant.

Table 1 shows that a number of significant observations can be made when comparing the performance of the freely suspended titania with the microcomposites. At acidic conditions, the photodegradation by the microcomposites and the free titania show similar rate constants and variation with the concentration of titania. In contrast, near a neutral pH the photocatalytic degradation of MO using the microcomposites has much smaller rate constant and remains largely unaffected even when the concentration of titania is increased. A closer look at the structure of the microcomposite can give some insight into these differences. The PAAc chains that are present as interpenetrating chains[8] in the microcomposite contain carboxylic acid groups that are mostly deprotonated above pH 4 and known to functionalize inorganic oxide surfaces[12-14]. Near a neutral pH, photocatalytic degradation of MO using the microcomposites is less because the negatively charged carboxylate groups interact with the oxide surface and can disrupt the adsorption of negatively charged hydroxyl species onto the titania surface due to electrostatic repulsion. Since the number of oxidative species is diminished, the lower photocatalytic oxidation reaction is experimentally manifested as a lower reaction rate constant. However, at pH 2 the PAAc is protonated and the titania surface within the microcomposite remains primarily unhindered. As a result, the photocatalytic performance of both the microcomposites and the freely suspended titania are comparable under acidic conditions.

Increasing the overall TiO_2 concentration in the solution from 50 to 200 ppm requires increasing the weight fraction of microcomposites in the solution. The resulting increase in polymer fraction has little influence at acidic conditions and the primary effect from increased titania surface sites dominates leading to faster photocatalysis. In contrast, at neutral pH conditions, the increase in the PAAc fraction balances the increase in titania and little change in the reaction rate constant can be observed as TiO_2 content changes from 50 ppm to 200 ppm.

CONCLUSIONS

The photocatalytic behavior of novel polymer-titania microcomposites that show rapid sedimentation in aqueous dispersions was studied. Using the photodegradation of a model organic dye, methyl orange, the photocatalytic behavior of the microcomposites was compared with freely suspended titania over a range of pH values. Under acidic conditions, the reaction rates were found to be identical while at a pH of 6.5, the freely suspended titania showed faster rate kinetics. However, at both pH values the microcomposites showed rapid sedimentation with settling velocities nearly 100 times faster than the freely suspended titania, which makes them promising candidates for applications such as wastewater remediation where the use of nanoparticles of TiO_2 is advantageous for photocatalysis but separation of the nanoparticles is difficult and time consuming.

ACKNOWLEDGEMENT

Financial support in the form of a graduate teaching assistantship from a NSF grant on Curriculum Reform (EEC-0530444) to CAC is gratefully acknowledged. The authors would also like to thank Bijith D. Mānkidy for help with photocatalytic experimentation.

REFERENCES

1.   M. D. Earle, Physical Review **61**, 56-62 (1942).
2.   A. Fujishima, T. N. Rao and D. A. Tryk, Journal of Photochemistry and Photobiology, C: Photochemistry Reviews **1** (1), 1-21 (2000).
3.   N. Serpone and R. F. Khairutdinov, Studies in Surface Science and Catalysis **103**, 417-444 (1997).
4.   M. R. Hoffmann, S. T. Martin, W. Choi and D. W. Bahnemann, Chemical Reviews **95** (1), 69-96 (1995).
5.   H. Chun, W. Yizhong and T. Hongxiao, Applied Catalysis, B: Environmental **30** (3,4), 277-285 (2001).
6.   S. A. Lee, K. H. Choo, C. H. Lee, H. I. Lee, T. Hyeon, W. Choi and H. H. Kwon, Industrial & Engineering Chemistry Research **40** (7), 1712-1719 (2001).
7.   R. L. Pozzo, J. L. Giombi, M. A. Baltanas and A. E. Cassano, Catalysis Today **62** (2-3), 175-187 (2000).
8.   C. A. Coutinho, R. K. Harrinauth and V. K. Gupta, Colloids and Surfaces, A: Physicochemical and Engineering Aspects **318** (1-3), 111-121 (2008).
9.   R. R. Bacsa and J. Kiwi, Applied Catalysis, B: Environmental **16** (1), 19-29 (1998).
10.   V. K. Gupta, R. Jain, A. Mittal, M. Mathur and S. Sikarwar, Journal of colloid and interface science **309** (2), 464-469 (2007).
11.   S. K. Kansal, M. Singh and D. Sud, Journal of hazardous materials **141** (3), 581-590 (2007).
12.   F. Zhu, J. Zhang, Z. Yang, Y. Guo, H. Li and Y. Zhang, Physica E: Low-Dimensional Systems & Nanostructures **27** (4), 457-461 (2005).
13.   M. Kurihara and H. Matsuyama, JP Patent 2004067740 Patent No. JP Patent 2004067740 (2004).
14.   V. Gupta, A. Kumar, C. Coutinho and S. Mudhivarthi, WO 2008052216

Mater. Res. Soc. Symp. Proc. Vol. 1171 © 2009 Materials Research Society   1171-S04-08-Q05-08

TiO$_2$ Anatase Nanotubes for the Purification of Uranium, Arsenic and Lead Containing Water: An X-ray Photoelectron Spectroscopy Study

Marco Bonato[1], K. Vala Ragnarsdottir[2,3] and Geoffrey C. Allen[1]
[1]Interface Analysis Centre, University of Bristol, Bristol, BS8 8BS, U.K.
[2]Department of Earth Sciences, University of Bristol, Bristol, BS8 1RJ, U.K.
[3]Institute of Earth Science, School of Engineering and Natural Sciences, University of Iceland, Hjardarhagi 6, Reykjavik 107, Iceland

ABSTRACT

TiO$_2$ anatase nanotubes synthesised via anodic oxidation were used as adsorbent for the uptake of U and Pb from aqueous solution and the photoremoval of As(III). An X-ray photoelectron spectroscopy study of the sorbent medium surface revealed a high adsorption of U and Pb at pH 8. The adsorption of the uranyl ion was enhanced in an anoxy (N$_2$) atmosphere, because this prevents the formation of very stable carbonyl complexes. As(III) was adsorbed on TiO$_2$ but in the presence of O$_2$ and UV light was oxidized to As(V). XPS analysis revealed that in the pH range 3-9 As(V) was always the major species detected at the surface of the titania photocatalyst.

INTRODUCTION

Titanium dioxide is one of the most important photocatalytic materials for ground water purification and heterogeneous photocatalysis. It offers an attractive low temperature, low cost, catalytic technique, for the decomposition and removal of organic pollutants in both the liquid and gaseous phase in the presence of UV light [1]. This surface reactivity makes it an excellent adsorbent material for the potential uptake of inorganic pollutants such as uranium, lead and arsenic from ground waters. In the case of arsenite (As(III), AsO$_3^{3-}$), it is possible to couple the adsorbent capability of titania with its photocatalytic properties. As(III) is easily adsorbed and oxidized on the surface of the titania to the arsenate ion, As(V) – AsO$_4^{3-}$, which is less noxious than the trivalent ion and easier to eliminate using coagulation methods [2,3]. In this communication we report a study of the adsorbtion of the metals uranium, lead and arsenic on TiO$_2$ using X-ray photoelectron spectroscopy. The amount of metal adsorbed was quantified by measurements of surface coverage expressed in atomic percent. TiO$_2$ anatase nanotubes have recently been given special scientific attention for their increased surface area, enhanced photocatalytic effect and simplicity of synthesis. The nanotubes used in this experimental work were synthesised via anodic oxidation of titanium in a fluorine bearing solution[4,5].

EXPERIMENTS AND METHODS

Synthesis and characterization of the TiO$_2$ anatase nanotube.

A titanium foil (1x1 cm) was mechanically polished under ambient laboratory conditions with increasingly fine grades of silicon carbide paper to 4000 grade (~2 μm). Preparation of the anatase nanotubes on the titanium metal surface was achieved by anodic oxidation at room temperature (21 °C), carried out for 20 min. at a constant 20 V, in 0.5 wt% HF aqueous solution with a platinum counter electrode. High spatial resolution images of the sample surface were recorded using scanning electron microscopy using a SEM Hitachi S-2300 and transmission electron microscopy with a Jeol 2010 TEM. The images shown in Fig. 2 revealed that a tubular TiO_2 array grew on the surface of the Ti metal precursor, separated by a thin (~20 nm) oxide layer. The nanotubes were about 300 nm in length with outer diameter 150 nm and wall thickness between 20-30 nm. The samples were then annealed at 320 °C for 5 h to induce transformation of the amorphous nanotubes into the desired anatase form. The resultant crystallinity of the nanotube surfaces was confirmed by Raman spectroscopy.

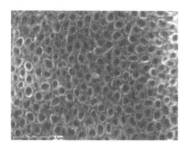 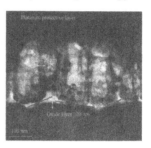

Figure 1. Images of anodic TiO_2 nanotube layers formed in 0.5 wt% HF at 20 V for 20 min; SEM top view (left) and TEM cross section (right).

Uranium and lead adsorption experiments.

In a typical experiment, a coupon of TiO_2 anatase nanotube was exposed for 48 hours at 20 °C to a 30 ml solution of 10 ppm uranyl acetate ($UO_2(CH_3COOH)_2 \times 3H_2O$) or 100 ppm of lead nitrate ($Pb(NO_3)_2$), in individual polyethylene screw cap vials. The pH was adjusted in the range 3-9 by adding 1M NaOH or 1M HCl, in solutions equilibrated in air; for alkaline solutions NaOH was added slowly, to prevent the formation of hydrocarbonyl complexes. Adsorption experiments with the uranyl ion were repeated in anoxic (N_2) atmosphere, where the solution was purged from O_2 and CO_2 by bubbling N_2 gas during the pH adjustment and for the duration of the experiment. After reaction, each coupon was removed from solution and rinsed in deionised water, and dried before analysis under high vacuum using XPS.

Arsenite photoadsorption experiments.

A 100 ppm arsenite stock solution ($NaAsO_2$, British Drug House Ltd, London UK) was formulated by dissolving prescribed quantities of sodium arsenite in deionised H_2O with gentle magnetic agitation to aid dissolution; the pH adjustments were performed as described above. The TiO_2 sample coupons were immersed in 50 ml of the As(III) solution at 20 °C for 24 h, in individual screwcap vials. Different samples were left to react in the total absence of light, then transferred with the solution into pyrex beakers. In a further experiment the system was held under direct UV light illumination ($\lambda = 360$ nm, 2 x 40 W) for 30 min and agitated by bubbling the solution with air. After reaction each coupon was removed from solution, rinsed in deionised water, dried and placed in the XPS instrument for analysis.

XPS measurements.

XPS spectra were acquired using a Kratos XSAM800 spectrometer with Al Kα X-rays (1486.6 eV; 140 W; 7 kV, 20mA, the vacuum was maintained at ~5 x 10^{-8} mbar). The instrument was interfaced with the software Pisces (Dayta Systems Ltd). The electron analyser was normal to the surface. Acquisitions were setup with 30 eV pass energy, 500 ms dwell times and 0.1 eV step size for high resolution scans, and 100 eV pass energy, 300 ms dwell times and 1 eV step size for wide scan spectra. The data set of a single acquisition regarded high resolution scans of the U4f core level photoelectron region from 378-405 eV, the Pb4f region from 135-150 eV, the As3d region 42-50 eV, the Ti2p core level region from 452-468 eV and the C1s region from 276-292 eV. The surface area analysed by XPS was about 10x10 mm. All the spectra were calibrated to the adventitious hydrocarbon C1s line at 285.0 eV. For the peak fitting a Shirley background subtraction was applied to all the high resolution spectra. Quantitative analysis of the elements detected was performed by measuring the absolute peak area after applying the relevant sensitivity factors[6]. The amount of metal (U or Pb) adsorbed was quantified by measurements of surface coverage expressed in atomic percent relative to the Ti2p photoelectron peak. The surface coverage of the adsorbate (M = U, Pb or As) at the surface of the adsorbent was expressed as percentage of the quantitative area A_M of the metal towards the total XPS peak area of the analysed peak ($A_{Tot} = A_M + A_{Ti}$). For arsenic, the oxidation state of the metal adsorbed was determined using curve fitting of the As3d photoelectron lines. The peak was resolved into the As(III) component and the oxidised species As(V), based on the binding energies of arsenic reported in the literature.

RESULTS AND DISCUSSION

XPS results of uranium adsorption.

The XPS spectra recorded from an anatase nanotube sample after reaction with uranyl acetate solution indicated the presence of sorbed uranium at the surface of the TiO_2 nanotube. The primary U4f peaks ascribed to the adsorbed metal were recorded at 382.2 eV for U4f$_{7/2}$ and 393.0 eV for U4f$_{5/2}$. No change in the oxidation state of the uranyl ion initially in solution was detected. The results are in good agreement with previous studies for uranium adsorption on titania[5]. The surface coverage of U on TiO_2 at different pH for the solution of uranyl acetate in air is plotted in Fig. 2. The sorption of the UO_2^{2+} ion increased with increasing pH, with a maximum at pH 7 (low pH sorption edge). Above pH 8 a sharp decrease in adsorption was observed, because at high pH there was a consistent presence of the very stable uranyl carbonate complexes $(UO_2)CO_3(OH)^-$, $UO_2(CO_3)_2^{2-}$ and $UO_2(CO_3)_3^{4-}$, formed from dissolved CO_2 in solution; These complexes deplete the uranyl ion from the solution, preventing any further adsorption of uranium on TiO_2. To eliminate dissolved CO_2, the sorption experiments were repeated purging the uranyl ion solution with N_2. Fig. 2 shows the surface coverage of U when the uranyl ion was adsorbed in the absence of CO_2. From the XPS data it was also possible to demonstrate that in the absence of carbonyl complexes the low pH sorption edge was shifted to higher pH value and the uptake of the uranyl ion enhanced. At more alkaline pH the adsorption fell. In this case the pH of the solution was higher than the point of zero charge for TiO_2 (5.8) and an electrostatic repulsion between the negatively charged TiO_2 surface and the uranyl hydroxide complexes explains the lower uranium uptake.

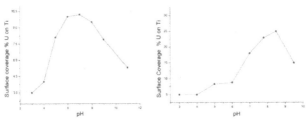

Figure 2. The pH dependence of the U(VI) sorption on TiO_2 nanotube expressed as surface coverage after the XPS analysis of the titania surface. Adsorption preformed in air (left) and in a nitrogen atmosphere (right).

XPS result of lead adsorption.

XPS analysis of the adsorbed lead showed the presence of a doublet centred at 137.9 and 143.8 eV, ascribed to the $Pb4f_{7/2}$ and $Pb4f_{5/2}$ peaks of Pb(II)[7]. The deconvolution of the XPS Pb4f peak did not show a characteristic change in the oxidation state of the adsorbed Pb. The XPS surface coverage for a 100 ppm lead nitrate solution studied in the pH range 3-9 is shown in Fig. 3. The adsorption of Pb at the surface of the TiO_2 nanotube surface increased with the pH, reaching a maximum at pH 8. Pb(II) sorption was a minimum at low pH due to lower electrostatic interaction between the dominant cationic Pb^{2+} and the positively charged titania surface. There was a sorption edge from pH 3, with a maximum at pH 7-8, that coinciding with pH values higher than the point of zero charge for TiO_2, when the negatively charged surface interacted with the mono-, di- and tetra-valent lead species ($Pb(OH)^+$, $Pb_3(OH)_4^{2+}$ and $Pb_6(OH)_8^{4+}$), as predicted by Abdel-Samad and Watson in their study of the adsorption of lead on goethite[8].

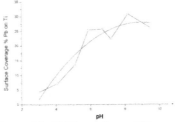

Figure 3. The pH dependence of the Pb(II) sorption on TiO_2 nanotube expressed as surface coverage for XPS analysis of the titania surface.

XPS results for arsenic adsorption.

The XPS investigation of the TiO$_2$ nanotube surface after its exposure to an arsenite solution under UV light irradiation determined the valence state of the arsenic ion adsorbed and oxidized at the surface of the TiO$_2$ catalyst. Due to the low XPS signal intensity of the As doublet peak (3d$_{5/2}$ and 3d$_{3/2}$), it was decided to deconvolute the signal as one unique 3d peak. As(III) and As(V) spectra were clearly identified at binding energy 44.5 and 45.8 eV respectively[9]. The percentage of oxidized As(V) and As(III) adsorbed on TiO$_2$ nanotubes after UV irradiation in the studied pH range is plotted in Fig. 4. It can be seen that the amount of the oxidized As(V) detected was always higher than the initially adsorbed As(III) and that the highest percentage of As(V) was recorded at the acidic and basic edges of the experimental pH range. As pointed out by Dutta *et al.*[2], for arsenate (AsO$_4^{3-}$) the highest adsorption is achieved at acidic pH. Indeed when pH was lower that point of zero charge of titania, the surface of the adsorbent medium was positively charged, so there existed an attraction between the positive surface sites of TiO$_2$ and the neutral or negatively charged As(V) species, AsO(OH)$_3$, AsO$_2$(OH)$^{2-}$, AsO$_3$(OH)$^{2-}$ and AsO$_4^{3-}$ (see ref. [2] and equations therein). With increased pH, As(V) species were once again more abundant than As(III) (Fig. 4). This contrasts with the previous study[2], hence As(III) was expected to have higher affinity with TiO$_2$ at alkaline pH. In alkaline solution the release of protons form As(OH)$_3$ removes hydroxyl groups coordinated at the surface of TiO$_2$, creating positively charged sites for the adsorption of As(III) anions. In our case the initial affinity for As(III) towards the TiO$_2$ surface, favored the adsorption and therefore the photooxidation of the initially present As(III), when the solution with the TiO$_2$ anatase nanotube catalyst was irradiated under UV light in oxygenated atmosphere. In this, As(V) was the major species detected by XPS measurements because the majority of As(III) was adsorbed and easily oxidized to As(V).

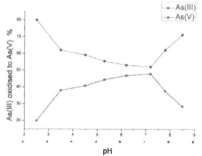

Figure 4. The pH dependence As(V) and As(III) distribution at the surface of TiO$_2$ nanotube after XPS analysis of a sodium arsenite 100 ppm solution irradiated with UV light.

CONCLUSIONS

In this paper we report the behavior of a TiO_2 anatase nanotube surface as an efficient adsorbent for the uptake of uranium and lead from water. XPS analysis of the surface of the adsorbent titania medium showed that the adsorption of both metals was higher at pH 7-8. In the case of uranium, the uptake was enhanced when the uranyl adsorption experiments were performed in a N_2 atmosphere. TiO_2 nanotubes were used as a photocatalyst for the oxidative removal of arsenite. In the pH range 3-9 and in the presence of UV light, the oxidized As(V) ion was found to be the major species detected at the surface of the catalyst, as illustrated by XPS analysis.

ACKNOWLEDGMENTS

The authors would like to acknowledge support (to MB) from the Marie Curie Early Stage Training Programme (MEST-CT-2005-020828) MISSION of the European Commission during the course of this work

REFERENCES

1. M. Arabatzis, S. Antonaraki T. Stergiopoulos , A. Hiskia, E. Papaconstantinou, M. C. Bernard and P. Falaras, *J. Photochem. Photobiol.*, **149**, 237-245 (2002).
2. P. K. Dutta, A. K. Ray, V. K. Sharma, F. J. Millero, *J. Colloid. Interface Sci.*, **278**, 270-275 (2004).
3. M.A. Ferguson, M. R. Hoffmann, J. G. Hering, *Environ. Sci. Technol.*, **29**, 1880-1886 (2005).
4. D.A. Gong , C. Grimes , O. K. Varghese, Z. Chen, W. Hu, E. C. Dickey, *J. Mater. Res.*, **16**, 3331-3334 (2001).
5. M. Bonato, G. C. Allen, T. B. Scott, *Micro&Nano Letters,* **3**, 57-61 (2008).
6. C. D. Wagner, L. E. Davis, M. V. Zeller and J. A. Taylor, R. H. Raymond and L. H. Gale, *Surf. Interf. Anal.*, **3**, 211-225, (1981).
7. T. Yoshida, T. Yamaguchi, Y. Iida and S. Nakayama, *J. Nucl. Sci. Technol.*, **40**, 672-678, (2003).
8. H. Abdel-Samad and P.R. Watson, *Appl. Surf. Sci.*, **136**, 46-54 (1998).
9. H.W. Nesbitt, I.J. Muir and A.R. Pratt, *Geochim. Cosmochim Acta*, 59, 1773-1786 (1995).

Mater. Res. Soc. Symp. Proc. Vol. 1171 © 2009 Materials Research Society   1171-S05-04

Conduction Band Edge of (Ti,Sn)O$_2$ Solid Mixtures Tuning for Photoelectrochemical Applications

J. Simiyu[*], B. O. Aduda and J. M. Mwabora,

[1]University of Nairobi, Dept. of Physics, P.O. Box 30197 - 00100, Nairobi, Kenya

ABSTRACT

We report investigation of effect of conduction band edge on the dye injection and transport by preparation of (Ti,Sn)O$_2$ solid mixtures in ratios of 80:20 and 90:10 as possible applications in dye sensitized solar cells. SEM micrographs showed highly porous with nanometer sized particles of around 6 - 10μm diameter. X-ray diffraction patterns showed strong TiO$_2$ anatase peaks with crystal orientation directions (101) being the strongest in both the solid mixtures and in pure TiO$_2$. XPS studies have shown an apparent chemical shift for Ti 2p and O1s core level spectra with an energy difference between the unmodified and the solid mixture being 0.65eV. Initial I-V studies have shown high open circuit potential (V$_{oc}$) but low short circuit photocurrent, showing a possible unfavorable band edge shift between the semiconductor and the dye LUMO level.

INTRODUCTION

Photoelectrochemical solar cells with wide-band-gap oxide semiconductors have received much attention since the development of dye-sensitized solar cell with porous TiO$_2$ thin film by Gratzel's group [1, 2]. The chemical and physical processes involved in the operation of these cells take place in two-phase system consisting of porous TiO$_2$ film and I$_3^-$/I$^-$ redox electrolyte. Charge injection from the photoexcited dye and regeneration of the dye by electron transfer from I$^-$ lead to transport of electrons in the TiO$_2$ as well as transport of I$_3^-$ and I$^-$ ions in the electrolyte with electron transfer from I$^-$ to the oxidized dye and regeneration of I$^-$ from I$_3^-$ at the counter electrode linking the two transport processes.

To achieve a favorable electron injection efficiency requires proper matching of the dye molecule's LUMO (Lowest Unoccupied Molecular Orbital) level (in excited state) with the conduction band of the semiconductor. Various approaches have been applied, among them using dye molecules that have high LUMO levels in excited state, modifying the conduction band edge of the semiconductor to either lower or raise it [3 - 6] and using redox couples with high molecular orbital levels while maintaining the same band edge for the semiconductor. These studies have been done extensively on single semiconductors like TiO$_2$, and to lesser extent ZnO and SnO$_2$ [7, 8].

Recently dye sensitized nanoporous semiconductor materials comprising more than one material have been studied as a possible way to improve on performance [9 – 12]. Tennakone et al. [13] reported the suppression of charge recombination for the mixture (referred to as composite) of SnO$_2$ with small crystalline size of 15nm and ZnO with large crystalline size of 2 mm. Tai et al [14 - 16] studied widely SnO$_2$-TiO$_2$ coupled and composite solar cells using various sensitization dyes. They have reported higher values of Incident Photon to Current conversion Efficiency (IPCE) in the coupled system compared with that of the composite system, attributing it to a better charge separation due to easy electron transfer in two semiconductor layers with different energy levels. These studies have been mainly on rutile type mixtures; however, anatase (Ti,Sn)O$_2$ solid mixtures have been reported before [17] but as potential applications in photocatalysis. An investigation of electron transport properties of these mixtures is of great importance for potential application in photovoltaics.

In the present study, sol–gel derived anatase nanostructured (Ti,Sn)O$_2$ solid mixtures were prepared at two ratios (80:20 and 90:10). The intention was to have SnO$_2$ formed inside TiO$_2$ hence only affecting the conduction band edge while maintaining crystal phase of the main semiconductor. Structural properties (SEM, ESCA & XRD) and electrochemical behavior were studied for the two ratios. Current voltage characteristics and electron transport studies were carried out on these solar cells.

EXPERIMENTAL DETAILS

The starting materials were titanium iso-propoxide (C$_{12}$H$_{28}$O$_4$Ti), tin chloride precursors (SnCl$_4$.5H$_2$O), iso-propanol and 0.1M ammonia solution. All reagents were of analytical grade and were supplied by Sigma Aldrich unless stated otherwise. 7.4g of SnCl$_4$.5H$_2$O was dissolved in 50ml iso-propanol then mixed with 60g titanium iso-propoxide. The resulting solution was then mixed drop wise slowly in 300ml of 0.1M NH$_3$ under heavy stirring. This method produced instant precipitates, which was then heated to 80°C and peptized at that temperature for 8 hours. The precipitate was then cooled to room temperature and 250ml portion measured out and autoclaved at a temperature of 220°C for 12 hours. This resulted in (Ti,Sn)O$_2$ suspension of 6.22%wt which was further concentrated in rotor vapor to 20%wt concentration. Finally the paste was centrifuged and washed in ethanol three times to remove the salts and iso-propanol to produce (Ti,Sn)O$_2$ colloidal solution containing 40%wt. The (Ti,Sn)O$_2$ colloid in ethanol was then mixed with 20.0g of tarpineol and further underwent stirring – sonication – stirring process. The same process was repeated after addition of 10% ethyl cellulose (Fluka GmbH, Germany) in ethanol to the colloid and then concentrated in a rotor vapor and collected in a reagent bottle. The resulting paste was about 21%wt of (Ti,Sn)O$_2$ solid mixture.

The above paste was used to coat films on F:SnO$_2$ conducting glass substrate by screen printing method. The mesh size was 6x8mm and one frame had 20 mesh openings that gave a total of 20 films at a single coating. To coat subsequent layers, the coated film was heated at a temperature of 150°C for five minutes and left to cool to room temperature then another coat applied until the required number of coatings was achieved. The films were then sintered in air for one hour at 450°C and left to cool to room temperature. These films were either used for XRD, SEM and ESCA analysis or assembled into complete solar cells for I-V, and transport/recombination analysis, after determining the film thickness. All films used in this study had 3μm film thickness.

X-Ray diffraction was obtained by means of Siemens D5000 Diffractometer (Bruker AXS GmbH, Germany) with θ-2θ Parallel beam geometry. The range was from 10° to 80° with detector type of scan at a scan speed of 0.6°/min and step size of 0.01°. (Ti,Sn)O$_2$ films for XRD analysis were coated on plain glass such that the SnO$_2$ peaks observed are ascribed to the presence of SnO$_2$ in the solid mixture. SEM micrographs were measured with LEO 1550 scanning electron microscope at 20kv electron source. ESCA spectra were recorded with

Quantum 2000 scanning ESCA probe (Physical Electronics Inc., USA) with Al K_α x-ray source with photon energy of 1486.6eV.

Screen printed films were cut to a size of 3 x 2cm and heat treated at a temperature of 300°C for one hour to remove chemisorbed water and left to cool to 80°C and then subsequently immersed in a dye bath consisting of 0.5mM ruthenium N719 dye complex in ethanol for 12 hours sensitization. Platinised counter electrodes were prepared by coating clean, predrilled conducting glass substrates (measuring 1.5 x 1.5cm), with 48mM of hydrochloroplatinic acid and heating the counter electrode at 450°C for 30 minutes in air and gradually cooled to room temperature. The dye sensitized (Ti,Sn)O_2 photoelectrode and counter electrode were then separated by a 25μm thick thermoplastic surlyin film (DuPont, USA), and sealed by heating. The internal space having the dye sensitized nanoporous film was filled with a drop of the electrolyte through the two holes on the counter electrode. The electrolyte consisted of 0.1M iodine, 0.1M LI, 0.6M TBA, 0.5M 4-terBPY in 3-methoxypropionitrile solvent medium. The electrolyte introduction holes were then sealed with another surlyn film under a thin glass cover by heating. Finally, silver was painted along the edges of the cell to make contacts and to reduce series resistance in the device.

I-V characteristics were obtained using Newport solar simulator model 91160 (Oriel Instruments, USA) to give an irradiation of 100mW/cm^2 (equivalent of one sun at AM 1.5) at the surface of the solar cells. All the solar cells characterized in this work had an active area of 0.48cm^2. The current – voltage characteristics under these conditions were obtained by applying an external potential bias to the cell and measuring the generated current with a Keithley 2400 digital source meter (Keithley, USA). This process is fully automated using LabVIEW^{TM} software from National Instruments Inc, USA.

The set up for electron transport and lifetime measurements consisted of Stanford Research Systems lock-in amplifier model SR 570 (Stanford Research Systems, USA), the HP 33120 function generator, and BNC 2110 data acquisition board (National instruments Inc, USA) controlled automatically by LabVIEW^{TM} (National Instruments Inc, USA) application. The light source was from a laser diode (LabLaser, Coherent LabLaser, USA) with λ_{max} 635nm. Switching from potentiostatic (to measure in short circuit) and galvanostatic (in open circuit) was achieved by use of a solid state switch. Electron transport and lifetime measurement followed a method developed by Boschloo et al. [18] for time resolved measurements. A small square wave modulation (<10% intensity, 0.1 – 2Hz) was added to the base light intensity. The solar cell response was fitted to exponential rise or decay function. Traces were averaged 10times for transport and 2 times for recombination studies.

RESULTS AND DISCUSSION

Structural Properties

From SEM analysis (Figure 1), the films produced were highly porous with nanometer sized particles. However, samples for solid mixtures ((b) and (c)) had slightly smaller particle size than for pure TiO$_2$ (d). This is possibly due to two main reasons, one being that the starting materials for both batches were different, i.e. pure TiO$_2$ samples were made from P25 Degussa which has large particle sizes while the solid mixtures were made from titanium isopropoxide and the colloid particle size was determined by the autoclave temperature (which was 220°C). Another possibility is that the presence of SnO_2 particles in the solid mixture may inhibit the growth of the colloid particles during autoclave digestion.

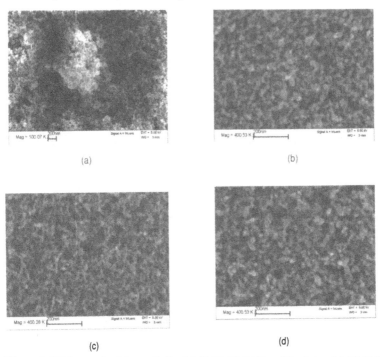

Figure 1: SEM scans for (a) pure SnO_2, (b) 80:20 $(Ti,Sn)O_2$ solid mixture (c) 90:10 $(Ti,Sn)O_2$ solid mixture and (d) TiO_2

This is because TiO_2 colloids prepared from titanium isopropoxide (not shown in this work) have typical sizes ranging from 10nm to 30nm in diameter depending on the autoclave temperature. An increase in autoclave temperature to 250°C did not yield any meaningful rise in the particle size. From XRD analysis' particle size determination using Scherer's formula, pure SnO_2 and solid mixtures gave particle sizes ranging from 6nm to 10nm. The size seemed not to affect dye sensitization as the solid mixtures showed very high dye attaching ability during sensitization. However, solar cells made from solid mixtures had very low I_{sc} which does not agree with high dye attaching ability (this will be discussed in detail under IV results later in the report).

From XRD scans (Figure 2), strong anatase peaks from TiO_2 were observed in solid mixture and in pure TiO_2 with crystal orientation directions (101) being the strongest. Previous works on solid mixtures have mainly produced rutile mixture, but in this work, solid mixtures in anatase form have been produced. Another aspect worth noting is that TiO_2 peaks in pure form do not show any peak shift in mixtures which suggests the absence of crystal phase change when SnO_2

and TiO_2 are mixed. This is important because the only property in focus in solid mixtures is the variation of conduction band edge that should only affect the dye injection and not phase change that may affect the matching between Ti and Sn and hence the electron transfer during transport.

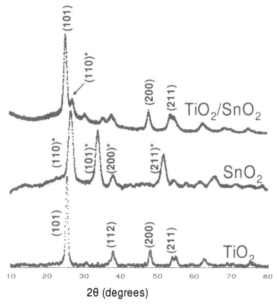

Figure 2: XRD surface scans for pure $(Ti,Sn)O_2$ and 80:20 $(Ti,Sn)O_2$ solid mixtures. Bragg reflections with * are for SnO_2 while the ones without are for TiO_2.

However, at this point, it is not conclusively established that the solid mixture formed is a case where Sn forms within Ti shell and both form solid mixture oxide. XPS analysis so far carried out does not also show that. In Figure 2 for the case of mixtures, the strongest peak for SnO_2 is in (101) direction but due to the small percentage composition compared to TiO_2 it appears to be very small.

141

ESCA Results

From XPS studies (Figure 3), there is an apparent chemical shift for Ti 2p and O1s core level spectra (figures (c) and (d) on one hand and (e) and (f) on the other); with the energy difference between the unmodified and the solid mixture is found to be 0.65eV. Chemical shifts in SnO_2 have been observed before [19], and have been attributed to possible change in Fermi level position in the semiconductor band gap or just due to band bending. This observation may be a similar occurrence in this case for Ti, which is very likely because it is also expected that O1s peaks shift with the same magnitude. From figures 3 (e) and (f), the energy shift is found to be 0.74eV which is within the range as of Ti 2p.

An interesting observation here is the absence of chemical shift in Sn 3d when mixed with TiO_2 (Figures 3 (a) and (b)). This may partly confirm that SnO_2 forms inside TiO_2 shell which may only affect the conduction band edge while maintaining crystal phase of the main semiconductor (in this case TiO_2). This property was earlier observed in XRD analysis in terms of phase shift.

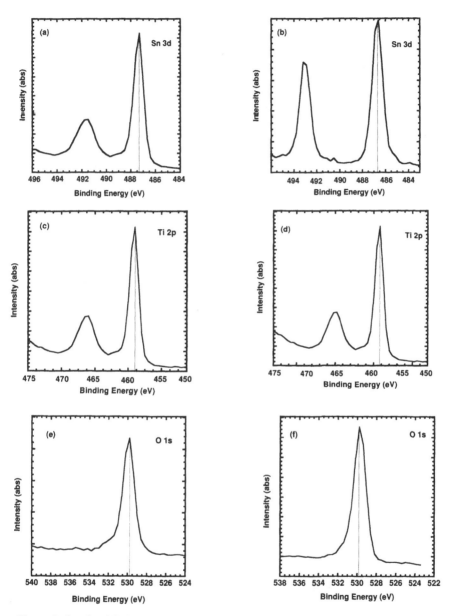

Figure 3: Core level XPS scans for (a) Sn 3d in pure SnO_2 (b) Sn 3d in 80:20 $(Ti,Sn)O_2$ mixture (c) Ti 2p in pure TiO_2 (d) Ti 2p in 80:20 $(Ti,Sn)O_2$ mixture (e) O 1s in TiO_2 and (f) O 1s in 80:20 $(Ti,Sn)O_2$ mixtures

I-V Characteristics

Figure 4 shows I-V characteristics for $(Ti,Sn)O_2$ solid mixture solar cells. Solar cell samples from this paste exhibited high Photovoltage but low photocurrent densities. The films showed very high dye adsorption during dye-sensitization, a property which is expected to translate to high electron harvesting and hence high photocurrent density; however that is not the case. This could therefore imply that the introduction of SnO_2 in the solid mixture could have raised the semiconductor conduction band edge instead of lowering it, making it hard for electrons to climb the energy barrier that is created. Further investigations are ongoing to ascertain the band edge position involving optical and I-V characterization. A change of the redox couple's concentration from 0.1M LI & I to 0.5M showed an improvement in the photocurrent recorded (plot (b)). This could further indicate that the raise in concentration caused a shift in the conduction band more positive than when 0.1M was used.

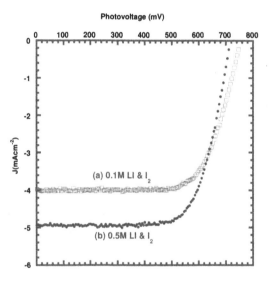

Figure 4: I-V characteristics of $(TiO_2Sn)O_2$ solid mixture films with different redox couple concentration

Electron Transport & Recombination Studies

Figure 5 shows electron transport and lifetime results for the solid mixture with 0.5M LI & I_2 redox electrolyte couple; (a) shows the current transient obtained for the sample and (b) is the transport time obtained from (a) plotted against the photocurrent at different diode light intensities.

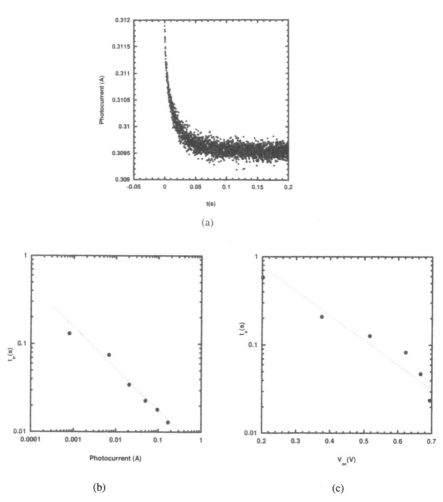

(a)

(b)

(c)

Figure 5. (a) Current transients for charge transport measurements, (b) transport time vs photocurrent and (c) electron lifetime vs photovoltage

The measured electron transport time (t_{tr}) and lifetime (t_e) are shown as function of photocurrent (b) and photovoltage (c) respectively at various light intensities. An exponential relationship is obtained between t_e as reported before in other studies [18, 20, 21]. Samples exhibited long t_{tr} times indicating slow transport process in the solid mixtures. One possible explanation for this is the inconsistent solid mixture formation which may lead to grain boundaries in the nanoporous structure. Another possible cause could be the formation of trap states in the solid mixtures. In general, electron lifetimes were longer than the transport times.

CONCLUSIONS

Sol-gel route has successfully been used here to produce titanium mixed oxide with anatase structure. XRD and ESCA studies have confirmed that mixed $(Ti,Sn)O_2$ was formed as a solid solution. It has been observed in this work that the particle size of the mixed oxide was smaller than that for the TiO_2 obtained in a similar way. This has been observed before by Marcela et al [17] and from Leite and coworkers' [19] report on a novel approach to control particle size of SnO_2 prepared by the polymeric precursor method, using Nb_2O_5 as dopant. This effect was attributed to surface effect of the precursor, an effect that may be occurring in the present work, i.e presence of SnO_2 restricting the growth to the surface.

Initial results have indicated that the use of 0.5M LI and 0.5M I_2 raised the photocurrent density indicating that there was a shift in conduction band more positive than when 0.1M was used. This could have improved on electron injection in the system hence the increased photocurrent density. However, ESCA results have not conclusively shown that indeed solid mixture was formed with SnO_2 forming inside TiO_2, this leaves room for further work on this and also explore other methods of preparation of this mixtures. More important in the preparation step will be the stage at which precipitates form and when the mixtures are formed.

ACKNOWLEDGEMENTS

This research was performed at Angstrom Laboratory, Uppsala University, Sweden. Authors thank MSL, for SEM and ESCA, Material Chemistry for the X-ray diffractograms, Anders Hagfeldt Group for discussions and hosting me, International Science Programs (ISP) for the fellowship and thanks also to the University of Nairobi for granting me a study leave during this research.

REFERENCES

[1]   B. O'Regan, M. Gratzel, Nature 353,737 (1991).
[2]   M.K. Nazeeruddin, A. Kay, I. Rodicio, R. Humphry-Baker, E. Muller, P. Liska, N. Vlachopoulos, M. Gratzel, J. Am. Chem. Soc. 115, 6382 (1993).
[3]   R. Cavicchi, M. Tarlov and S. Semanchik, J. Vacc. Sci. Tech., A8, 2347 (1990).
[4]   D.F. Cox, T.B., Fryberger and S. Semanchik, Surface Science, 224, 121(1989).
[5]   A. Ciera, A Cornet, J.R, Morante, S.M. Oilaizola, E. Castino and J Gracia, Mater. Sci. Eng. B, 66/70, 406 (2000).
[6]   J. Szuber, G. Czempic, R, Larcuprete and B Adamowicz, Sensors & Actuators B 65, 375 (2000).

[7]   S.E. Lindquist, A. Hagfeldt, S. Sodergren, H. Lindstrom, in: G. Hodes (Ed.), Wiley, Weinheim, pp. 169–200 (2001).

[8]   J.B. Asbury, E. Hao, Y. Wang, H.N. Gosh, T. Lain, J. Phys. Chem. B **105**,4545 (2001).

[9]   C. Nasr, P.V. Kamat, S. Hotchandani, J. Phys. Chem. **102**, 10047 (1998).

[10]   K. Tennakone, G.R.R.A. Kumara, I.R.M. Kottegoda, V.P.S. Perera, Chem. Commun. 1999 15 (1999).

[11]   K. Tennakone, J. Bandara, P.K.M. Bandaranayake, G.R.R.A. Kumara, A. Konno, Jpn. J. Appl. Phys. **40** L732 (2001)

[12]   G.R.R.A. Kumara, K. Tennakone, V.P.S. Perera, A. Konno, S. Kaneko, J. Phys. D: Appl. Phys. **34** 868 (2001).

[13]   K. Tennakone, P.K.M. Bandaranayake, P.V.V. Jayaweera, A. Konno, G.R.R.A. Kumara, Physica E **14** 190 – 196 (2002)

[14]   Weon-Pil, Tai, Kozo Inoue, Eosin Y-sensitized nanostructured SnO_2/TiO_2 solar cells, Materials Letters **57** 1508–1513 (2003).

[15]   Weon-Pil Tai, Solar Energy Materials & Solar Cells **76** 65–73 (2003)

[16]   Weon-Pil Tai, Kozo Inoue, Jae-Hee Oh, Solar Energy Materials & Solar Cells **71** 553–557 (2002)

[17]   M. Marcela, D. Oliveira, C. Danielle, Schnitzler, and A.J. G. Zarbin, Chem. Mater. **15**, 1903-1909 (2003)

[18]   G. Boschloo, L. Haggman, A. Hagfeldt, J. Phys Chem B 110, 13144 (2006)

[19]   E.R Laite, I.T. Weber, E. Longo, J.A. Varela, Adv. Mater. **12**, 965 (2000).

[20]   J. Nillsfolk, F. Kristofer, A. Hagfeldt, & G. Boschloo, J. Phys. Chem B, **110 (36)**, 17715 – 17718 (2006)

[21]   G. Schlichthorl, S. Huang, J. Sprague, A. Frank, J. Phys Chem B 101, 8141. (1997)

147

Mater. Res. Soc. Symp. Proc. Vol. 1171 © 2009 Materials Research Society   1171-S05-10

Photooxidation of Water Using Vertically Aligned Nanotube Arrays: A Comparative Study of TiO2, Fe2O3 and TaON Nanotubes

Mano Misra, Subarna Banerjee and Susanta K. Mohapatra
Materials Science and Engineering, University of Nevada, Reno, NV, 89557, U.S.A.

ABSTRACT

There is a real need for a material which absorbs in the visible light of the solar spectrum, is stable in water and at the same time economical. One-dimensional vertically aligned nanotubes have contributed to a great extent towards the visible light driven photoelectrolysis of water. In this work, we give an overview of the different nanotubes obtained through anodization of various metals and their application in the visible light photooxidation of water.

INTRODUCTION

Solar light driven water oxidation has received great attention due to the potential of technology for the production of green and renewable fuel.[1-4] The use of sunlight to produce the electrochemical potential required for electro-synthesis has been proposed as a potential means of harnessing solar energy long ago.[5] This photoelectrochemical (PEC) synthesis of hydrogen has an advantage over simple photocatalysis of water to evolve hydrogen (H_2). In a PEC system, the gases evolved (hydrogen and oxygen) are separated by compartmentalization of the system, thus eliminating the need for collecting the gases separately. The development of photocatalysts which use visible light efficiently is envisioned to be successful for this application.[6,7] Ordered nanotubes (NTs) formed by anodization of metals channelize the electrons formed by splitting of water through the back contact and thus give better performance than other one dimensional materials. Among all the nanotubes used for PEC generation of H_2, titania (TiO_2) nanotubes (NTs) prepared on a Ti foil by the electrochemical anodization method have attracted a lot of attention because of their potential to harvest sunlight and use the photoelectrons efficiently to generate H_2.[8-16] Because TiO_2 is a large band gap semiconductor (band gap = 3.1-3.2 eV) it absorbs only solar light in the UV region. The major problem is that only about 4-5% of the solar spectrum falls in this UV range. To improve the photocatalytic efficiency and efficient use of the solar spectrum, researchers have adopted different strategies such as changing the electrical properties of TiO_2 by varying the crystallite size;[17,18] doping TiO_2 with metal/nonmetal ions in order to induce red shift to the band gap.[19-22] It is predicted that the ideal band gap for a semiconductor to split water efficiency should be around 2.0 eV.[23] In this work, we report the synthesis, characterization and application of metal oxide and metal oxynitride NT arrays for PEC generation of hydrogen. The synthesis and application of hybrid TiO_2 NTs, ultra-thin iron oxide (Fe_2O_3) and tantalum oxynitride (TaON) NT arrays for visible light PEC hydrogen generation is discussed here.

EXPERIMENT

Materials. Ethylene glycol (Fischer, 99.5%), ammonium fluoride (NH_4F, Spectrum, 98%), titanium (Ti) foil (ESPI, 99.9% purity, 0.2 mm thick), iron (Fe) foil (Alfa Aesar, 99.99%, 0.25 mm thick), tantalum (Ta) foil (Sigma-Aldrich, 99.9%, 0.25 mm thickness) 1-Butyl-3-methylimidazolium tetraflouroborate (BMIM-BF$_4$, Fluka, 97%). All chemicals are used in the as-received condition without any further purification.

Synthesis of metal oxide nanotube arrays. Synthesis of metal (Ti, Fe, Ta) oxide nanotubes are carried out on a metal foil in organic media (3-10% water in ethylene glycol + 0.5 wt% NH$_4$F) using sonoelectrochemical anodization method (100 W, 42 KHZ, Branson 2510R-MT) under various conditions (Table I). The said processes are carried out using a two electrode system (flag shaped 1.0 cm^2 metal foil as anode and Pt foil, 3.75 cm^2 as cathode; the distance between cathode and anode is kept at 4.5 cm). The anodized samples are properly washed with distilled water to remove the occluded ions, dried in an air oven, and processed for annealing. A chemical vapor deposition furnace (CVD, FirstNano) furnace is used to convert the amorphous materials into crystalline one. TiO$_2$ nanotubes and Fe$_2$O$_3$ NTs are annealed under nitrogen and hydrogen (10% in Ar) respectively. Both the nanotubes are annealed at 500 °C for 3 h. The flow rate of the gases is maintained at 400 sccm (standard cubic centimeters per minute) at a heating and cooling rate of 1 °C/min. The annealed samples are then subjected to characterization and photoelectrochemical measurements.

Synthesis of TaON NTs. Anodization of Ta is carried out in a solution of ethylene glycol, 3 v% water, 0.5 wt% NH$_4$F at 100 V for 2 minutes. TaON NTs are made from as anodized Ta$_2$O$_5$ NTs by heating them in an atmosphere of flowing NH$_3$ (flow rate: 10 ml min^{-1}) at 700 °C for 6 h. Annealing is also done along with nitridation which reduces the defects in the material and also crystallizes it.

Characterization. A field emission scanning electron microscope (FESEM; Hitachi, S-4700) is used to analyze the nanotube formation and morphology. Diffuse reflectance ultraviolet and visible (DRUV–Vis) spectra of the samples are measured from the optical absorption spectra using a UV-Vis spectrophotometer (UV-2401 PC, Shimadzu). Fine BaSO$_4$ powder is used as a standard for baseline and the spectra are recorded in a range 200-800 nm. A scanning transmission electron microscope (STEM; Phillips CM 300) equipped with ESVision software is used for mapping and crystal distribution of the samples. A small amount of sample is placed in a carbon coated Cu-grid and subjected for High Resolution Transmission Electron Microscopy (HRTEM) measurement.

Photoelectrochemical tests. The experiments on H$_2$ generation from water are carried out in a glass cell with photoanode (metal oxide NTs/metal or metal oxynitride NTs/metal) and cathode (Pt foil) compartments. The compartments are connected by a fine porous glass frit. Silver/silver chloride (Ag/AgCl) is used as the reference electrode. The cell is provided with a 60 mm diameter quartz window for light incidence. The electrolyte used is an aqueous solution of 1 (M) potassium hydroxide (KOH). A computer-controlled potentiostat (SI 1286, England) is used to control the potential and record the photocurrent generated. A 300-W solar simulator (69911, Newport-Oriel Instruments, USA) is used as the light source. An AM 1.5 filter is used to obtain one sun intensity, which is illuminated on the photoanode (87 mW/cm^2, thermopile detector from Newport-Oriel is used for the measurements). The samples are anodically polarized at a scan rate of 5 mV/s under illumination and the photocurrent is recorded. All the experiments are carried out under ambient conditions.

Table I. Anodizing conditions of different metals to get the corresponding oxide NTs.

Material	Anodizing solution	pH	Anodization voltage (V)	Anodization time (min)
Single wall TiO$_2$ NTs	5v % H$_2$O+ EG+ 0.5 wt% NH$_4$F	6.6-6.8	60	60
Double wall TiO$_2$ NTs	10 v% H$_2$O +EG + 0.6 v% BMIM-BF$_4$	6.4	60	20
Fe$_2$O$_3$ NTs	3-5 v% H$_2$O +EG + 0.5 wt% NH$_4$F	6.6-6.8	50	13
Ta$_2$O$_5$ NTs	3 v% H$_2$O +EG + 0.5 wt% NH$_4$F	6.6-6.8	100	2

Ethylene glycol = EG

DISCUSSION

(1) Single wall TiO$_2$ NTs. Self-organized and vertically oriented TiO$_2$ nanotube arrays are synthesized on a Ti disc by sonoelectrochemical anodization method at 20 V for 1 h. The FESEM micrograph of annealed TiO$_2$ NTs showed individually separated nanopores of TiO$_2$ (Figure 1a). Figures 1(a) and (b) show the front and bottom views of The TiO$_2$ NTs respectively. The average internal diameter of the nanotubes is ~60 nm and the walls of the nanotubes are ~15 nm thick. Potentiodynamic plot of N$_2$ annealed single-wall NT arrays under global AM 1.5 illumination conditions show a current density of 0.64 mA/cm^2 (Figure 3A) Figure 3 shows the PEDC activity of both single and double wall TiO$_2$ NTs.

Figure 1. SEM images of TiO$_2$ NTs: (a) Front view and (b) bottom vie image of the NTs.

(2) Double wall TiO$_2$ NTs. When Ti is anodized in a solution that does not contain the simple fluoride anion (F-TiO$_2$ NTs), but the complex anion fluoroborate, double wall TiO$_2$ NTs are obtained (BF$_4$-TiO$_2$ NTs). The double-walls are formed in approximately 20 min of anodization at 60V (Figure 2). Each NT consists of concentric NTs with the average outer diameter of 82 ± 2 nm and 206 ± 4 nm. This is the largest ever diameter observed for NTs prepared by anodization process. These NTs are quite thick with average wall thickness of 27.5 nm and 49 ± 1 nm compared to single wall TiO$_2$ NTs (around 20 nm). Cross sectional SEM image showed that both the NTs have a single root (tube end). Unlike in the fluoride medium these NTs do not grow more than 350 nm under these experimental conditions, rather they form layer-by-layer structure on the Ti foil. Increasing the applied potential from 60 V to 80 V also did not affect the length of the NTs, rather it formed a multiporous NTs. A single NT is found to be made of a cluster of NTs due to multiple pitting on the nano-walls and in the barrier layer at higher applied potential. The N$_2$ annealed double-wall TiO$_2$ NTs have been tested for photoelectrolysis of water using global AM 1.5 and visible light sources. The N$_2$ annealed sample shows photocurrent density of 1.65 mA/cm^2 at 0.5 V$_{Ag/AgCl}$, which is 2.5 times higher than the nanotubes prepared using fluoride (0.64 mA/cm^2) (Figure 3A). To find the contribution of the visible light components on the total activity of the NTs, experiments are carried out with UV filters (only $\lambda \geq 400$ nm is

illuminated) (Figure 3B). The dark current is same in both the materials (3.7 µA) (Figure 3A). It is observed that unlike single-wall TiO_2 NTs, double-wall NTs possess good visible light photoactivity (18% compared to 0.39% using single-wall TiO_2 NTs) (Figure 3B). The better performance could be attributed to higher porous structure, better visible light absorption and 1–D architecture. From the above discussions, it can be summarized that NTs prepared using ionic liquid opens new opportunities for visible light driven water splitting reactions.

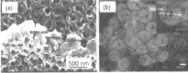

Figure 2. SEM image (a) and (b) STEM surface view of double-wall TiO_2 NTs.

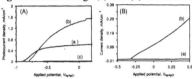

Figure 3. Potentiodynamic plot of annealed single-wall (a) and double-wall (b) NT arrays under the illumination of (A) global AM 1.5, and (B) $\lambda \geq 400$ nm. (c) Represents the dark current of both the materials. The electrolyte used is 1M KOH solution.

(3) Fe_2O_3 NTs. Higher potential (50 V) and an increase in water content helped to form self-standing Fe_2O_3 NTs with pore diameters of 50-55 nm, a wall thickness of 5-7 nm and a length ~3.7 µm (Figure 4 a). The as-anodized Fe_2O_3 NTs formed are amorphous in nature and are converted to crystalline structure after annealing at 500 °C in hydrogen atmosphere (Figure 4b) Anodization under lower potential only yielded porous structure. The NTs are found to be smooth and self-standing on the Fe foil. The formation of iron oxide structure is also observed from absorption studies (DRUV-Vis). The metallic colored Fe foil which does not absorb in the visible region turned to reddish-brown color after anodization, which significantly absorbed light up to 550 nm with the onset being around 629 nm (band gap = 1.97 eV). This is in accordance with the band gap of hematite. Figure 4(c) shows the PEC plot of Fe_2O_3 NTs. Figure 4(c) represents the PEC plot of Fe_2O_3 NTs. The maximum photocurrent density obtain under AM 1.5 conditions is 1.4 mA/cm^2. Under visible light conditions ($\lambda \geq 400$ nm), the current density is 0.7 mA/cm^2. The dark current The dark current density (without illumination) in all the cases is observed to be ~3 µA/cm^2 at 0.5 $V_{Ag/AgCl}$.

Figure 4. (a) FESEM image of Fe_2O_3 NTs, (b) TEM and FFT pattern of Fe_2O_3 NTs and (c) PEC plot of Fe_2O_3 NTs. Curve (a) in Figure (c) shows the photocurrent density under AM 1.5 conditions, curve (b) shows the photocurrent density under $\lambda \geq 400$ nm, curve (c) shows the dark current of Fe_2O_3 NTs.

(4) TaON NTs. Figure 5 (a) shows the FESEM images of the Ta$_2$O$_5$ NTs. The FESEM image of TaON NTs is identical with that of Ta$_2$O$_5$ NTs. FESEM images confirmed that the NTs are stable after nitridation. The tubes are having 50 ± 5 nm internal tube diameter and length 525 nm. The HRTEM image of TaON NTs shows that the NTs are highly crystalline having monoclinic structure with interplanar distance is 0.36 nm (Figure 5(b)). The FFT pattern also shows that the material is very crystalline. Under global AM 1.5 conditions, oxygen annealed TaON NTs showed a photocurrent density of 2.6 mA/cm^2 at 0.5 V$_{Ag/AgCl}$ (Figure 5(c)). The dark current is ~ 90 µA/cm^2. This indicates that the photocurrent obtained from the system is mostly due to the photocurrent generated by the illumination of light on the photoanode. The visible light activity of the TaON NTs photocatalyst is observed around 47%, which is quite significant compared to other nanotubes (Table II). The catalyst is run for 2 h and no significant change in photoactivity is observed. No significant changes are observed in the absorption intensity (DRUV-Vis) and surface morphology (SEM) of the composite photoanode after the sample has been used for photoelectrochemical tests. These results show that the composite photocatalyst is having potential for long term operation with good photoactivity.

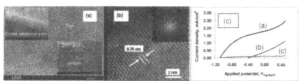

Figure 5. (a) FESEM images of Ta$_2$O$_5$ nanotube arrays on Ta foil, (b) HRTEM and FFT pattern of TaON NTs. This proposed process makes highly crystalline TaON NTs. (c) Potentiodynamic plot of TaON NTs under global AM 1.5 solar light (a), visible light (≥ 400 nm) illumination (b) and dark conditions.

Table II. Comparison of photocurrent density of TaON NTs with various other photocatalysts

Catalyst	Photocurrent density (mA/cm^2)[a] at 0.5 V$_{Ag/AgCl}$	Visible light contribution (%)[b]
P25/Ti	0.365	0.32
Single walled TiO$_2$ NTs/Ti	0.638	0.39
Double walled TiO$_2$ NTs/Ti	1.65	
Fe$_2$O$_3$ NTs/Fe	1.4	50
Fe$_2$O$_3$ nanoparticle/Fe	0.004	NA
Ta$_2$O$_5$ NTs/Ta	0.25	0.28
TaON NTs/Ta	2.6	47

[a] under AM 1.5 illumination conditions
[b] calculated using UV filter (Edmund optics)

CONCLUSIONS

In conclusion, it can be said that metal oxide NT arrays have been found to be the most efficient photocatalyst compared to other architectures. The visible light activity increased in the order Ta$_2$O$_5$<TiO$_2$<Fe$_2$O$_3$=TaON NTs. Further studies are going on to improve the performance

of these nanotubes as well as to develop other metal oxide NTs which may give more insight in the PEC generation of H_2 by photoelectrolysis of water.

REFERENCES

1. J. A. Turner, *Science* **285**, 687 (1999).
2. J. Nowotny, C. C. Sorrell, L. R. Sheppard, T. Bak, *Int. J. of Hydrogen Energy* **30**, 521 (2005).
3. P. V. Kamat, *J. Phys. Chem.* **111**, 2834 (2007) and references are cited therein.
4. S. Blanchette Jr., *Energy Policy* **36**, 522 (2008).
5. A. Fujishima, K. Honda, *Nature* **238** (5358), 37 (1972).
6. F. E. Osterloh, *Chem. Mater.* **20**, 35 (2008).
7. "Solar Hydrogen Generation" K.S. Rajeswar, R. McConnell, S. Litch, (Eds.), Springer, (2008).
8. G.K. Mor, K. Shankar, M. Paulose, O.K. Varghese, C.A. Grimes, *Nano Lett.* **5**, 191 (2005).
9. J.M. Macak, H. Tsuchiya, A. Ghicov, P. Schmuki, *Electrochem. Commun.* **7**, 1133 (2005).
10. J. H. Park, S. Kim, A.J. Bard, *Nano Lett.* **6**, 24 (2006).
11. K.S. Raja, M. Misra, V.K. Mahajan, T. Gandhi, P. Pillai, S.K. Mohapatra, *J. PowerSources* **161**, 1450 (2006).
12. J. Park, S. Bauer, K. von der Mark, P. Schmuki, *Nano Lett.* **7**, 1686 (2007).
13. S.K. Mohapatra, M. Misra, V.K. Mahajan, K.S. Raja, *J. Phys. Chem. C* **111**, 8677 (2007).
14. S.K. Mohapatra, M. Misra, *J. Phys. Chem. C* **111**, 11506 (2007).
15. S.P. Albu, A. Ghicov, J.M. Macak, R. Hahn, P. Schmuki, *Nano Lett.* **7**, 1286 (2007).
16. W. Chanmee, A. Watcharenwong, C. Chenthamarakshan, P. Kajitvichyanukul, N. R. de Tacconi, K. Rajeswar, *J. Am. Chem. Soc.* **130**, 965 (2008).
17. M. Ampo, T. Shima, S. Kodama, Y. Kubokawa, *J. Phys. Chem.* **91**, 4305 (1987).
18. Y. Xu, Z.Z. Zhu, W. Chen, G. Ma, *Chin. J. Appl. Chem.* **8**, 28 (1991).
19. K. Wilke, H.D. Breuer, *J. Photochem. Photobiol., A* **121**, 49 (1999).
20. U. Diebold, *Surf. Sci. Rep.* **48**, 53 (2003).
21. T.L. Thompson, J.T. Yates Jr., *Chem. Rev.* **106**, 4428 (2006).
22. H. Park, C.D. Vecitis, W. Choi, O.Weres, Hoffmann, M. R. *J. Phys. Chem. C* **112**, 885 (2008).
23. A.B. Murphy, P.R.F. Barnes, L.K. Randeniya, I.C. Plumb, I.E. Grey, M.D. Horne, J.A. Glasscock, *Int. J. Hydrogen Energy* **31**, 1999 (2006).

Mater. Res. Soc. Symp. Proc. Vol. 1171 © 2009 Materials Research Society   1171-S07-12

RF Sputter Deposition of Indium Oxide / Indium Iron Oxide Thin Films for Photoelectrochemical Hydrogen Production

William B. Ingler Jr. and Abbasali Naseem
University of Toledo, Department of Physics and Astronomy,
MS 111, McMaster Hall, Toledo, OH. 43606, U.S.A.

ABSTRACT

This project focuses on using indium oxide and indium iron oxide as an alloy to make a protective thin film (transparent, conductive, and corrosion resistant or TCCR) for amorphous silicon based solar cells, which are used in immersion-type photoelectrochemical cells for hydrogen production. From the work completed, the results indicate that samples made at 250 °C with 30 Watt of indium and 100 Watt of indium iron oxide, and a sputter deposition time of ninety minutes produced optimal results when deposited directly on single junction amorphous silicon solar cells. At 0.65 Volts, the best sample displays a maximum current density of 21.4 mA/cm^2.

INTRODUCTION

This project focuses on using indium oxide and indium iron oxide to make a protective thin film (transparent, conductive, and corrosion resistant, TCCR) to deposit on top of amorphous silicon based solar cells [1-4]. These TCCR coated amorphous silicon (a-Si) solar cells are used for hydrogen production in immersion-type photoelectrochemical (PEC) cells (Figure 1, Figure 2) The results indicate that samples made at 250 °C with 30 W of indium and 100 W of indium iron oxide sputter coated on to amorphous silicon solar cells are the optimal parameters for a stable TCCR coating alloying these two materials.

Indium oxide is an ideal layer in a hybrid photoelectrodes with amorphous silicon triple junction solar cells to function as TCCR layer. For a TCCR layer on a-Si solar cells, the standard conditions are the protective top layer needs to be at least 90% transparent; the layer need to be conductive at 7 mA/cm^2 or greater at 1.7 V (maximum power point voltage for a-Si); and it must be corrosion resistant for thousands of hours. Additionally, the top layer material typically needs to be a wide band gap material and should mismatch to the bands of the a-Si triple junction. However, pure In_2O_3 has rather poor stability in basic electrolyte so it needs to be alloyed to a compound that can stabilize the film, and the films need to be deposited at low temperatures (under 270 °C) so that the a-Si:H bonds will not degrade during In_2O_3-$InFe_2O_4$ deposition. These thin films were created using rf (radio frequency) sputter deposition. Two 2" sputter guns were used with respective targets of indium and indium iron oxide. The two main variables considered were the temperature of the depositions as well as the sputter powers of indium and indium iron oxide used. Characterization was done by current-voltage measurements and atomic force microscopy.

Indium oxide and indium iron oxide films were deposited using rf magnetron sputter deposition in argon and oxygen using multiple sputter guns simultaneously. The substrates used for this study included borosilicate glass, ITO glass, and triple junction a-Si solar cells on stainless steel. Substrate heating is achieved via two halogen lamps housed above the substrate with uniform heating created by rotation of the sample prior to and during deposition by a small mechanical motor. A 4" × 4" a-Si solar cell was used as a substrate and was loaded into a 4" × 4" substrate holder and placed inside the vacuum deposition chamber.

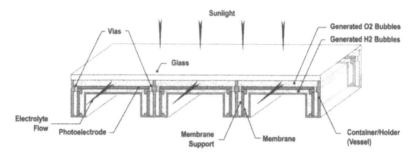

Figure 1. Schematic design for an immersion-type PEC. The photoelectrode stack is expanded in Figure 2.

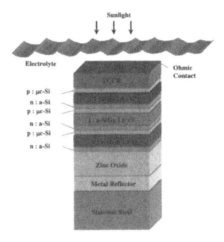

Figure 2. Schematic of an immersion-type PEC with a triple junction thin film silicon (tf-Si) photoelectrodes covered with a transparent, conductive, and corrosion resistant (TCCR) protection layer.

EXPERIMENT

Preparation of In_2O_3-$InFe_2O_4$ thin films

Indium oxide and indium iron oxide films were deposited using rf magnetron sputter deposition in argon and oxygen using multiple sputter guns simultaneously. The substrates used

156

for this study were single junction amorphous silicon (a-Si) solar cells deposited on 316 stainless steel substrate. Substrate heating is achieved via two halogen lamps housed above the substrate with uniform heating created by rotation of the sample prior to and during deposition by a small mechanical motor.

Rf power for the $InFe_2O_4$ target was varied at 50 to 150 W while the power for the In target was varied from 10 to 40 W. Substrate temperature was varied from 230 to 270 °C, and the oxygen concentration was held constant at 5% by volume in argon ambient. Two flows of gas were fed to the deposition chamber using mass flow controllers (MKS) with ranges of 0 to 10 sccm, one for pure argon and the other a mixture of 10% oxygen and the balance in argon. Chamber pressure was held at a constant 6 mTorr, and the deposition time was 60, 90, or 120 mins.

Measurements of Photocurrent

Photoelectrochemical measurements were made in a custom made vessel with a quartz window to reduce light scattering and absorption. The electrode set-up consisted of a working electrode (In_2O_3-$InFe_2O_4$ thin film), which had an average area of ~1.0 cm^2, a counter electrode (platinum gauze, Alfa Aesar), and a reference electrode (Radiometer Analytical saturated calomel electrode, SCE). The surface of the In_2O_3-$InFe_2O_4$ thin film electrode was illuminated with a 100 W xenon lamp (Oriel Instruments) with a light intensity of one sun. The intensity of the light was measured with a digital multimeter (Radioshack, model number 22-813) and a calibrated silicon detector (UDT Sensors, Inc.). The electrolyte solution for the n-type electrodes was 1.0 M potassium hydroxide (KOH). Photocurrent (mA/cm^2) as a function of applied potential (V) vs. saturated calomel electrode (SCE) were measured using a Voltalab PGZ 301 potentiostat and plotted using VoltaMaster 4 Electrochemical Software version 5.10. An applied potential range of −1.0 to +3.0 V was used at a scan rate of 20 mV/s.

For measurement purposes, a small portion of the 4" × 4" sample was cut out from the center region of the sample which was the most uniform portion of the sample. After cutting the sample, the edges were etched with 1 M hydrochloric acid to removed the shunted portion of the sample. Non-conductive epoxy was then coated over the etched surface with a little overlap over the remaining coated surface to cover exposed a-Si solar cell.

Atomic force microscopy (AFM) measurements

AFM measurements were done with a PicoSPMII system (Molecular Imaging).

DISCUSSION

Photocurrent Measurements

For the research on this paper we set the temperature at 250 °C, argon flow at 8 sccm, and argon with 10% oxygen at 2.67 sccm. The indium target power was varied from 10 to 40 W with an rf power supply and the indium iron oxide target was varied from 50 to 150 W using an rf power supply. In Figure 3 indium was sputtered at 20, 30, and 40 W with $InFe_2O_4$ at a constant 100 W. Dark current is measured by running a varied voltage through the semiconductor from -0.4 V to +1.0 V with no illumination on the semiconductor surface. All activity that is measured is produced from electrical activity only. Then the same surface si illuminated with a light source. Activity during illumination is caused from semiconductive and electrical behavior; however, when the dark current is subtracted from the illuminated current, the net result is photocurrent activity. Our figures used chopped scans where the light is put on the sample' for a set time and then removed for the same time to see if the recovery periods between light and dark are sharp which would indicate instantaneous movement of electrons in the semiconductor. From results shown in Figure 3, 30 W indium showed the best photocurrent response; however, the dark current is significant which means that the film crystalline structure is not fully formed which causes problems with forming stable film structures.

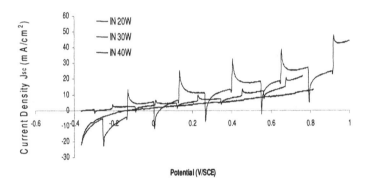

Figure 3. Varying the indium deposition power reaches a maximum at 30 W with 100 W $InFe_2O_4$ at 90 min deposition time.

In Figure 4, the total deposition time was investigated. Deposition times of 60, 90, and 120 minutes were done, but the 60 minute samples were shunted and the results are not shown. The 120 minute samples appear to be higher, but the dark current is greater which results in the photocurrent being equal to the 90 min sample. Figure 5 shows the stability scans for 30 W In and 100 W $InFe_2O_4$ where the sample was scanned from -1 V to +3 V forwards and reverse scans for three cycles. The current density is lower for this sample compared to the I-V data in Figures 3 and 4 since a different part of the larger sample was tested but partially shunted which affected the overall results. Table 1 summarizes the data for the variable deposition time trials.

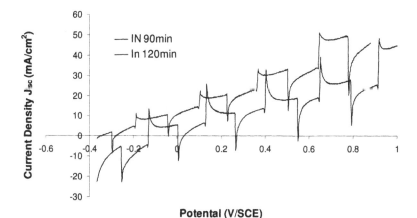

Figure 4. 30 W indium and 100 W InFe$_2$O$_4$ deposited at 90 and 120 min total deposition times. Note that the 120 min sample show higher total current density, but that the photocurrent density is equal since the dark current for these oxides is higher than normal when deposited at these low temperatures. Experimentation has shown that dark current goes to zero and current density increases with annealing at 500 °C for 2 hours.

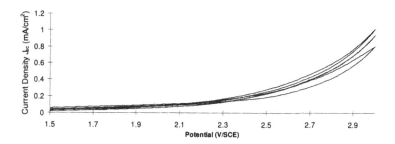

Figure 5. The stability plots for a sample made with 30 W indium and 100 W InFe$_2$O$_4$ at 90 min deposition time run from -1 V to +3 V for 3 cycles. After the initial sweep the I-V curve stabilizes.

Table 1. The corrosion rates for samples made with 30 W indium and 100 W InFe$_2$O$_4$ with deposition times of 60, 90, and 120 min.

Time	Corrosion Rates	In power	InFe$_2$O$_4$ power	Maximum Current Density (mA/cm^2)	Voltage at Maximum Current Density (V)
60 min	3.5 μm/year	30 W	100 W	0.0027	-0.068
90 min	9 μm/year	30 W	100 W	21.4375	0.648
120 min	6.2 μm/year	30 W	100 W	10.675	0.081

AFM Measurements

AFM images in Figure 6 show how the deposition conditions such as temperature affect the film quality. Good quality films show a smoother, less-detailed surface (Figure 6a). Meanwhile, films with no stability demonstrate a very rough, featured surface as shown in Figure 6b.

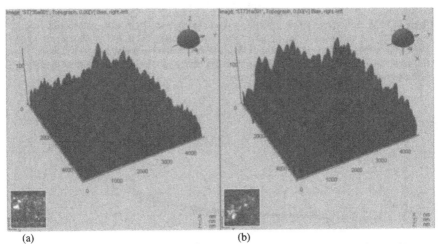

(a)   (b)

Figure 6. AFM images for (a) stable films demonstrate a uniform level surface, whereas (b) unstable surface show rough surfaces.

CONCLUSIONS

This study demonstrates our initial effort to deposit a TCCR material on a-Si solar cells. We did this deposition using 30 W indium, 100 W InFe$_2$O$_4$, and deposited for 90 min which demonstrated the best results to date. However, the films still need to be improved since the crystallinity is not ideal as the dark current is considerably high. Further studies will continue to focus on the total time of deposition to complete analysis with these two targets. The next step will be to deposit optimized TCCR coating on a triple junction solar cell and measure hydrogen production in a 4" × 4" immersion-type PEC module.

ACKNOWLEDGMENTS

This work was supported by sub-contract from Xunlight Corp. on a U.S. Department of Energy Grant DE-FG36-OSGO15028. The authors would like to thank Prof. Xianbo Liao and Shibin Zhang for producing single junction a-Si solar cells used on this project which were produced on a U.S. Department of Energy Grant DE-FG36-08GO18073.

REFERENCES

1. Ingler, W. B. Jr. and S. U. M. Khan, *Thin Sol. Films.* 2004. p. 301-308.
2. Ingler, W. B. Jr. and S. U. M. Khan, *Int. J. Hyd. Ener.* 2005. p. 821-827.
3. Ingler Jr., W.B.; Sporar, D.; Deng, X. "Sputter Deposition of In-Fe_2O_3 Films for Photoelectrochemical Hydrogen Production" *ECS Trans.* Vol. 3 (State-of-the-Art Program on Compound Semiconductors 45 (SOTAPOCS 45) -and- Wide Bandgap Semiconductor Materials and Devices 7), **2006**, 253.
4. Ingler Jr, W. B.; Ong, G.; Deng, X. "RF Sputter Deposition of Indium Oxide – Iron Oxide Films for Photoelectrochemical Hydrogen Production" *ECS Trans.* Vol. 16, No. 7. (State-of-the-Art Program on Compound Semiconductors 49 (SOTAPOCS 49), **2008**, 49.

Mater. Res. Soc. Symp. Proc. Vol. 1171 © 2009 Materials Research Society   1171-S07-14

Photocatalysis of nano-perovskites

Yen-Hua Chen, and Yu-De Chen
Department of Earth Sciences, National Cheng-Kung University, Tainan, Taiwan

ABSTRACT

In this study, we can successfully synthesize nano-perovskites, including nano-$CaTiO_3$, nano-$SrTiO_3$, and nano-$BaTiO_3$, by a co-precipitation method. The band gap of the nano-perovskites are 3.65 eV, 3.44 eV, and 3.35 eV, for nano-$CaTiO_3$, nano-$SrTiO_3$, and nano-$BaTiO_3$, respectively. The ability of photocatalysis for nano-$BaTiO_3$ is a little bit better than other nano-perovskites. It is also observed the photocatalytic activity increases with the increasing amount of photocatalysts. Moreover, the ability of photocatalysis using a higher energy UV-light is not promoted with the low energy UV-light.

INTRODUCTION

Very recently, the environmental pollution is more and more serious, such as water-pollution, air-pollution and so on. It is very important for researchers to find out the solution to the problem. The nano-perovskites have the catalysis and photocatalysis properties due to the strong redox and photocatalyzer [1-2]. For example, it can photocatalyze the toxic gas, organic materials, etc. [3-4].

In this study, nano-$BaTiO_3$ (90 nm/ round shape), nano-$CaTiO_3$ (80 nm/ rectangular morphology), and nano-$SrTiO_3$ (50 nm/ square form) are synthesized by a co-precipitation method [5-9]. We want to investigate the photocatalysis on organic materials using nano-perovskites, which is compared with nano-rutiles (TiO_2). In addition, the effects on the amount of nano-perovskites and the energy of UV-light for the photocatalytic property are also discussed.

EXPERIMENT

In this study, we synthesize nano-perovskites by the co-precipitation method, which is shown in Fig. 1. First, the chemical powders of $CaCO_3$, KOH, NaOH and TiO_2 are added into deionized water. When the materials are dissolved completely, it is then heated with a constant stirring rate. After that, it is calcined, and then cooled down to room temperature. We use the UV-Vis spectrometer to examine the band gap of the nano-perovskites. Moreover, the 254 nm and 365 nm UV-light are used to perform the photocatalysis experiment, respectively. The organic material is Methylene blue solution (M.B.). The photocatalysts are nano-$BaTiO_3$ (nano-BTO), nano-$CaTiO_3$ (nano-CTO), and nano-$SrTiO_3$ (nano-STO) with the amount of 0.03g and 0.1g. The experimental flowchart is shown in Fig. 2.

Fig. 1 The sketch of the co-precipitation method.

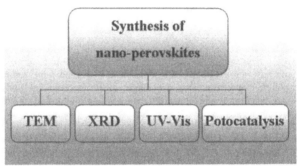

Fig. 2 The flowchart of the experiment.

DISCUSSION

Fig. 3 shows XRD patterns of nano-CaTiO$_3$, nano-BaTiO$_3$, and nano-SrTiO$_3$. Nano-CaTiO$_3$ has a orthorhombic structure; nano-SrTiO$_3$ and nano-BaTiO$_3$ are both cubic structure. From the XRD pattern, we can estimate the full width at half maximum (FWHM), which roughly represents the crystallinity of the specimens. The FWHM of nano-CaTiO$_3$ is around 0.22^0, and that of nano-BaTiO$_3$, and nano-SrTiO$_3$ are about 0.21^0 and 0.20^0, respectively. It indicates the sample of nano-SrTiO$_3$ probably has a better crystallinity. By the formula, $D = K * \lambda / \beta * \cos\theta$ (nm), we can calculate the particle size of nano-CaTiO$_3$ is 30~55nm, and that of nano-BaTiO$_3$ and nano-SrTiO$_3$ are about 30~55nm, and 40~45nm, respectively.

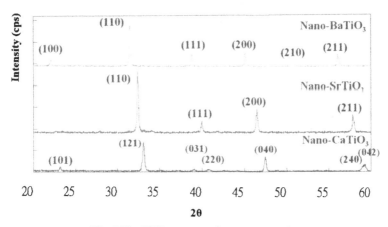

Fig. 3 The XRD patterns of nano-perovskites.

TEM images of nano-perovskites are presented in Fig. 4. The particle size distribution of nano-CaTiO$_3$ is 50~110nm; nano-SrTiO$_3$ is 40~100nm; nano-BaTiO$_3$ is 40~100nm. It indicates the size distribution is in the wide range. However, they are all in the nano-scale level. It is observed that the distribution of particle size is different from XRD results. It is because the XRD peaks are influenced by many factors, such as the grain size, the strain, and etc. Therefore, the crystal size measured from these two techniques are so different. However, we are convinced the TEM information because of the double check of the TEM observation.

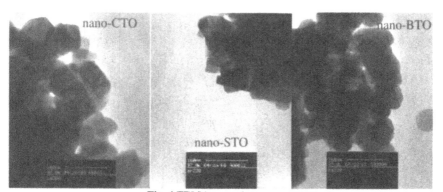

Fig. 4 TEM image of nano-perovskites.

The absorbance of UV-Vis spectrum for the nano-perovskites is shown in Fig. 5. From the result, we can calculate the band gap of the nano-perovskites. It is 3.65 eV for nano-CaTiO$_3$, 3.44 eV for nano-SrTiO$_3$, and 3.35 eV for nano-BaTiO$_3$.

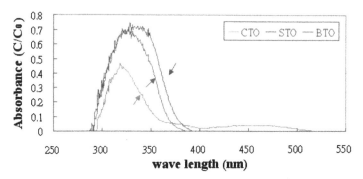

Fig. 5 The UV-Vis spectrum of nano-perovskites.

Fig.6 (a) and (b) show the photocatalysis property of nano-perovskites. It is observed the ability of photocatalysis for nano-perovskites with 0.03g is very poor. When the amount of photocatalysts increases to 0.1g, the photocatalytic activity becomes apparent. Among them, nano-BaTiO$_3$ has the better photocatalytic ability, however, they are all worse than nano-rutiles.

(a)

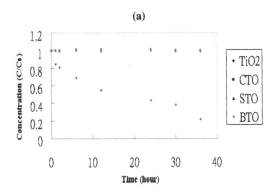

(b)

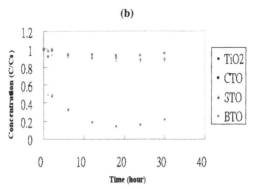

Fig. 6 The photocatalysis on M.B. for nano-perovskites with (a) 0.03g, (b) 0.1g. The wavelength of UV-light is 365 nm.

Fig. 7 shows the ability of photocatalysis of nano-perovskites. The UV-light has a wavelength of 254nm. It is found the concentration of M.B. does not change a lot with the higher energy of UV-light. The ability of photocatalysis for all nano-perovskites is not good.

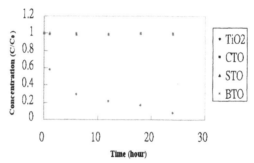

Fig. 7 The variation of M.B. concentration photocatalyzed by the nano-perovskites. The wavelength of UV-light is 254 nm.

CONCLUSIONS

We can successfully synthesize nano-perovskites, including nano-CTO, nano-STO, and nano-BTO, by a co-precipitation method. The band gap of all the nano-perovskites are in the wavelength belong to UV-light. The ability of photocatalysis for nano-BTO is a little bit better than other nano-perovskites, although they are all worse than nano-rutiles. It is also observed the ability of photocatalysis increases with the increasing amount of the photocatalyst. Moreover, the ability of photocatalysis using the high energy UV-light is not enhanced. The previous photocatalytic studies about simple perovskites usually focus on nano-STO. Form our experiment, however, nano-BTO has the better photocatalytic activity than that of nano-STO. In order to promote the photocatalysis properties, we can dope some elements into the nano-BTO or reduce the particle sizes of nano-BTO. This will be on going in the future.

REFERENCES

1. Liu J.W., Chen G., Li Z.H., and Zhang Z.G., International Journal of Hydrogen Energy, 32, 2269 (2007)
2. Lei X.F., and Xue X.X., Materials Chemistry and Physics, 112, 928 (2007)
3. Liu Y., Xie L., Li Y., Yang R., Qu J., Li Y., and Li X., Journal of Power Sources, 183, 701 (2008)
4. Mizoguchi H., Ueda K., Orita M., Moon S.C., Kajihara K., Hirano M., and Hosono H., Materials Research Bulletin, 37, 2401 (2008)
5. Qi J.Q., Wang Y., Chen W. P., Li L.T., and Chan H.L.W., Journal of Solid State Chemistry, 178, 279 (2008)
6. Lee S., Son T., Yun J., Kwon H., Messing G.L., and Jun B., Materials Letters, 58, 2932 (2004)
7. Chen J.F., Shen Z.G., Liu F.T., Liu X.L., and Yun J., Scripta Materialia, 49, 509 (2003)
8. Holliday S., and Stanishevsky A., Surface & Coatings Technology, 188, 741 (2004)
9. Miao J., Hu C., Liu H., and Xiong Y., Materials Letters, 62, 235 (2007)

Mater. Res. Soc. Symp. Proc. Vol. 1171 © 2009 Materials Research Society   1171-S07-19

Photocatalysis for the Degradation of Ionic Surfactants in Water: The Case of DPC

Alberto Naldoni[a], Claudia L. Bianchi[a], Silvia Ardizzone[a], Giuseppe Cappelletti[a], Luca Ciceri[a], Alessandro Schibuola[a], Carlo Pirola[a] and Marco Pappini[b]

[a]Dipartimento di Chimica Fisica ed Elettrochimica, Università degli Studi di Milano, Via Golgi 19, 20133 Milan, Italy
[b]C.I.G.A., Via Golgi 19, 20133 Milan, Italy

ABSTRACT

Traditional techniques to remove contaminants (carbon adsorption, incineration, biological activity and chemical treatment) have a lot of disadvantages. Advanced Oxidation Processes (AOP's) are used as alternative processes in the degradation of surfactants and in general for wastewater treatment. They are based on the generation of OH·, one of the most powerful oxidant known ($E° = 2.73$ V) and capable to react non-selectively with any organic compound. In the present work, the degradation of a cationic surfactant (dodecylpyridinium chloride (DPC)) was performed. The photodegradation reaction was investigated both in a slurry reactor and in a vessel where the photocatalyst (P25 by Degussa) was anchored onto an aluminum surface to avoid the final filtration of the powder at the end of the reaction. Moreover a new photoreactor was built on purpose to investigate the influence of the pressure on the degradation process.

INTRODUCTION

Surfactants are wetting agents that lower the surface tension of a liquid, allowing easier spreading, and lower the interfacial tension between two liquids. The surfactants present in detergent products remain chemically unchanged during the washing process and they are discharged down the drain with the dirty wash water. In the vast majority of cases, the drain is connected to a sewer and ultimately to a waste water treatment plant, where the surfactants present in the sewage can be removed by biological and physical-chemical processes. European Law now requires efficient treatment of urban waste water and all but the smallest conurbations would have been complied before the end of 2005 [1].

Environmental remediation of surface water containing detergents, dyes and pesticides is particularly complex also due to the necessity of avoiding biodegradation processes especially when cationic surfactants are present. Consequently, the literature is rich in studies which show different approaches (called advanced oxidation processes) for the degradation of such molecules.

The most valuable and attractive ability to degradate and to mineralize completely or nearly so, organic pollutants is at the very basis of photocatalytic reactions applied to environmental remediation efforts. Many studies have been published on the use of TiO_2 as a photocatalyst for the decomposition of organic compounds [2].

Many authors have already investigated the degradation of ionic surfactants by photocatalysis, especially anionic molecules, and their conclusion was that the photodegradation of the surfactant depends on many factors such as: pH, flow rate, loading of photocatalyst, and the presence and quantity of extraneous oxidative agents [3].

In this study the degradation of a cationic surfactant (DPC, dodecylpyridinium chloride) is performed by means photocatalysis using commercial TiO_2 samples both in a slurry reactor and in a vessel where the photocatalyst was anchored onto an aluminum surface to avoid the final filtration of the powder at the end of the reaction. Moreover a new photoreactor was built on purpose to investigate the influence of the pressure on the degradation process. The realized apparatus is cylindrical, to allow homogeneous irradiation and prevent any shadow regions, and can operate between 1 and 20 bar. It has a double-walled cooling system with an inner part (1.3 L) where the reaction takes place. It is equipped with a magnetic stirrer at the bottom of the reactor. It is made of steel AISI 316 with the irradiation lamp directly in the centre of the main body. A quartz candle, transparent to the UV radiation, was used to protect the lamp from the pressure.

From an analytical point of view, spectrophotometric and TOC analyses were performed to investigate both the degradation of the starting molecule and the final degree of mineralization of the overall process.

In the case of the photodegradation treatment at ambient pressure, the degradation path was studied in detail by means of mass spectrometry.

EXPERIMENT

DPC, dodecylpyridinium chloride, was purchased from Aldrich (purity > 99%). Degussa P25 titanium dioxide was employed as photocatalyst.

All degradation runs were carried out employing the experimental setup similar to that already described [5]. It consisted of a cylindrical, tightly closed Pyrex vessel (500 mL), which was irradiated by external lamps. At the beginning of the runs the reactor was always almost completely filled up with the aqueous solution or suspension containing the surfactant, in order to minimize the headspace of the reactor and thus any loss of the volatile substrate. All runs were carried out at (30±5) °C under continuous stirring at ca. 220 rpm. One external iron halogen lamp (Jelosil, model HG 500), emitting in the 310–400 nm wavelength range, was employed as irradiation sources. The effective power consumption, measured amperometrically, was 500W.

Aqueous solutions or suspensions initially contained a fixed DPC concentration of 0.5 mM or 0.1 mM. In photocatalytic runs, 0.1 g L^{-1} of titanium dioxide was added to the solutions directly in the reactor. Samples (3 mL) were withdrawn from the reactor at different reaction times during the runs, through a rubber septum on the reactor cover, and analyzed spectrophotometrically in a Perkin-Elmer Lambda 16 apparatus. Prior to analysis, TiO_2 was separated from the suspensions by centrifugation at 4800 rpm for 15 min. The extent of mineralization was determined through total organic carbon (TOC) analysis using a Shimadzu TOC-5000A analyzer. The maximum duration of the runs was 6 h.

A proper reactor was also realized to allow the investigation of the influence of pressure on both the photolysis and photocatalytic process (Fig.1). A powerful irradiation source (125 W) was used in the present setup, emitting in the wavelength range of 254-364 nm.

The realized apparatus is cylindrical, to allow an homogeneous irradiation and prevent shadow region, and can operate between 1 and 20 bar. It is equipped with a magnetic stirrer at the bottom of the reactor and is made of steel AISI 316 with the irradiation lamp directly in the centre of the main body, set vertically in the middle of the reactor. A quartz candle, transparent to the UV radiation, was used to protect the lamp from the pressure.

Fig 1_Set-up of photocatalytic-pressure reactor.

DISCUSSION

Fig.2 shows the comparison between DPC degradation in TiO_2 slurry and deposited onto an Al lamina, respectively. Large differences in the degradation performance can be observed by both changing the adopted procedure and increasing the amount of pollutant in solution. In experiments performed with catalyst in slurry a high DPC degradation was obtained. Passing from DPC concentration of 0.5 mM to 0.1 mM a 60 % increase of molecule degradation with complete disappearance at the end of reaction was observed. Instead, by anchoring titanium dioxide on the aluminum surface (foil 1: 8cm x 11cm, foil 2: 8cm x 4cm), a maximum of 30 % degradation was obtained. Both catalyst loading and surface area exposed to surfactant solution affect the degradation yield (fig. 2), even if the most important factor remains surfactant concentration in solution.

(a)   (b)

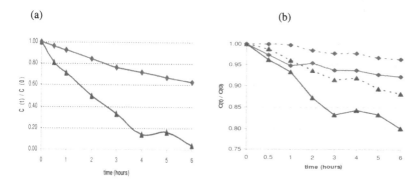

Fig.2_ Comparison between DPC disappearance: *catalyst in slurry* (a) red line [DPC] = 0.5 mM, blue line = [DPC] = 0.1 mM; *anchored catalyst* (b) red line [DPC] = 0.5 mM and TiO_2 = 0.1 g L^{-1}, blue line [DPC] = 0.1 mM and TiO_2 = 0.3 g L^{-1}, solid line (foil 2), dotted line (foil 1).

In order to obtain indications concerning the photodegradation path samples withdrawn from the reactor at intermediate reaction times were analyzed by High Resolution Mass Spectrometry.

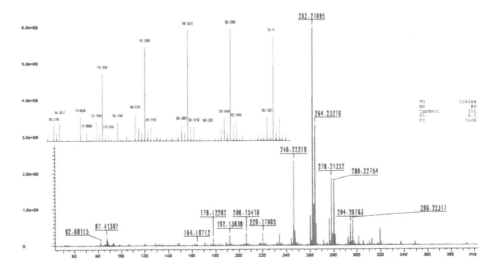

Fig. 3_ High Resolution Mass Spectra of DPC 0.5 mM at 120 min. The inset is a zoom of lower m/Z value than m/Z of DPC (248.23796).

From mass spectra it possible to suppose that two different types of OH radicals may occur. From the higher mass/charge (m/Z) than DPC m/Z it's clear that the OH radical randomly attach to the aromatic ring and also to the alkylic chain. From lower m/Z species we supposed that a decarboxylation mechanism occur over the alkylic chain passing through different carbonyl intermediates (aldehyde and/or ketone, carboxylic acid, beta ketoacid) as shown briefly in Fig. 4:

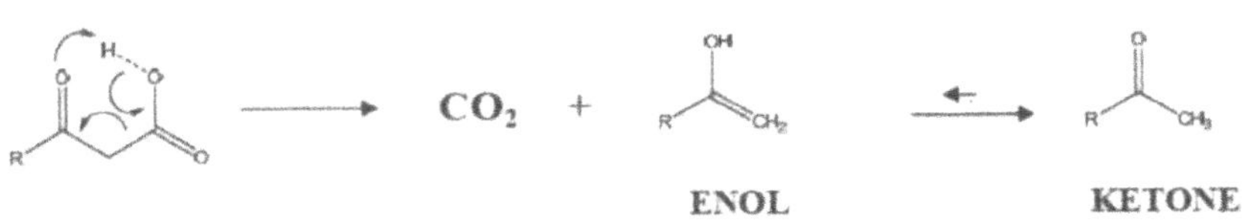

$$CO_2 +$$

ENOL   KETONE

Fig. 4_ Decarboxylation mechanism that occur over alkylic chain of DPC.

The pressure influence was firstly investigated with experiments performed in the dark. No surfactant degradation was observed, not even at 15 bar pointing out the absolute stability of the DPC molecule even at high pressure. An immersion UV source (125 W) was used in the present study, emitting in the wavelength range of 254-364 nm. As DPC is able to absorb part of the emitted light, its direct photolytic degradations in the absence of any photocatalyst was first investigated kinetically, to be compared with that of its photocatalytic degradation in the presence of P25. As shown in Fig. 5, direct photolysis produced appreciable surfactants degradation after a 6 h-long irradiation and this degradation is highly facilitated by the pressure. In particular, it is possible observe a linear increase of the mineralization degree with increasing the working pressure up to a 90% of DPC at 15 bar. This result is particularly interesting due to the usual toughness to degrade cationic surfactants.

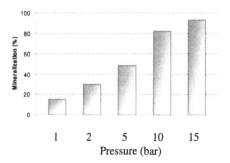

Fig.5_Mineralization of DPC by photolysis under several different pressures.

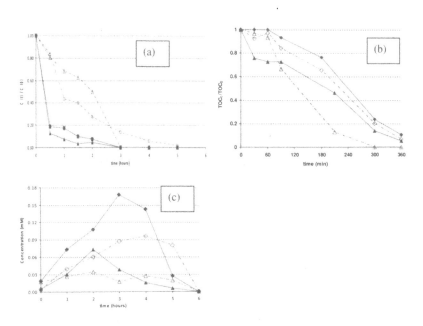

Fig.6_Degradation of DPC by both photolysis (solid lines) and photocatalysis (dotted lines) at 15 bar: (a) DPC disappearance at λ_{max} = 278 nm; (b) DPC mineralization investigated with total organic carbon analysis; (c) Aldehydes concentration profile. Red line [DPC] = 0.5 mM, blue line [DPC] = 0.1 mM.

Since with photolysis experiments at 15 bar the best percentage of DPC mineralization, was obtained, phocatalysis at that pressure was investigated in detail (Fig. 6a-b-c). In Fig. 6a and 4b the comparison between photolysis and photocatalysis at different concentration is reported. It is possible to observe that photocatalysis is the faster procedure only at 0.1mM of DPC mineralization experiment (Fig 6b) while in DPC disappearance experiment (Fig 6a) simple photolysis is more efficient.

The formation of aldehyde as intermediates during the degradation, was investigated by spectrophotometric analysis (Nash reagent method) [6] (Fig.6c). Using different experimental conditions photocatalysis and photolysis it was observed a temporal shift in maximum aldehyde peak that is an indication that different degradation path was occurred. It is possible to observe a fast increase of the aldehydes concentration into the treated solution up to 3 hours and then a decrease till their complete disappearance confirming that these species are really intermediates of the overall reaction, as also confirmed by High Resolution Mass Spectrometry.

CONCLUSIONS

In conclusion, with the present study it is possible to highlight different important points:

- At room pressure fully satisfying DPC degradation/mineralization with TiO_2 in slurry.
- Great influence of working pressure (>2 bar) on direct DPC photolysis.
- At 15 bar enhancement of the photocatalytic DPC degradation/mineralization.
- Good correlation between experimental data (mass spectroscopy vs. aldehyde test) to represent degradation mechanism.

REFERENCES
[1] Council Directive 91/271/EEC of 21 May 1991 concerning urban waste water treatment.
[2] S. Ardizzone, C.L. Bianchi, G. Cappelletti, S. Gialanella, C. Pirola, V. Ragaini, J. Phys. Chem. C, 111 (2007) 13222
[3] T. Zhang et al., Appl. Catal. B: Env. 42 (2003) 13
[4] E. Selli, C.L. Bianchi, C. Pirola, M. Bertelli, Ultrason. Sonochem. 12 (2005) 395–400.
[5] C.L. Bianchi, C. Pirola, V. Ragaini, E. Selli, Appl. Catal. B: Environ. 64 (2006) 131–138.
[6] Nash T., Biochem. J., 55, 416 (1953).

Mater. Res. Soc. Symp. Proc. Vol. 1171 © 2009 Materials Research Society   1171-S09-03

THE U.S. DEPARTMENT OF ENERGY'S WORKING GROUP ON PHOTOELECTROCHEMICAL HYDROGEN PRODUCTION: PROMOTING TECHNOLOGY-ENABLING BREAKTHROUGHS IN SEMICONDUCTOR MATERIALS RESEARCH

R. Garland
Office of Hydrogen, Fuel Cells and Infrastructure Technologies
Office of Energy Efficiency and Renewable Energy
U.S. Department of Energy
Washington, DC 20585

E. L. Miller
Hawaii Natural Energy Institute
University of Hawaii at Manoa
1680 East-West Road POST 109, Honolulu, Hawaii 96822

ABSTRACT

Photoelectrochemical (PEC) hydrogen production, using sunlight to split water, is an important enabling technology for a future "Green" economy which will rely, in part, on hydrogen as an energy currency. The traditional semiconductor-based PEC material systems studied to date, however, have been unable to meet all the performance, durability and cost requirements for practical hydrogen production. Technology-enabling breakthroughs are needed in the development of new, advanced materials systems, and toward this end, the U.S. Department of Energy's Working Group on PEC Hydrogen Production is bringing together experts in analysis, theory, synthesis and characterization from the academic, industry and national-laboratory research sectors. Key Working Group activities, as described in this paper, include performing techno-economic analyses of large-scale PEC production systems and establishing standardized testing and screening protocols for candidate PEC materials systems. In addition, a number of Working Group "Task Forces" are focused on advancing critical PEC materials theory, synthesis and characterization capabilities for application in the research and development of broad-ranging materials systems of promise, including complex metal-oxide and -nitride compounds, amorphous silicon alloys, III-V semiconductors and the copper chalcopyrites. The current status of Working Group activities and progress is summarized.

INTRODUCTION

Photoelectrochemical (PEC) hydrogen production, the splitting of water into hydrogen and oxygen using sunlight, is an important enabling technology for a future Green Economy which will rely, in part, on hydrogen as an energy currency [1]. The traditional semiconductor-based PEC material systems studied to date, however, have been unable to meet all the performance, durability and cost requirements for practical hydrogen production. For example, PEC semiconductors such as titanium-dioxide and other metal-oxides have proven to be stable in aqueous solutions, but suffer from low solar conversion performance due to their high band gaps [2, 3]. Based on these inherent limitations, it is becoming increasingly clear that new, more advanced materials will need to be developed. Technology enabling breakthroughs in materials R&D are needed for the success of PEC hydrogen production.

Toward this end, the U.S. Department of Energy (DOE) currently funds a number of research institutions from the academic, industrial and national laboratory sectors with the objective of discovering, engineering and optimizing such advanced PEC materials systems for solar water-splitting. Both the Energy Efficiency and Renewable Energy (EERE) and Basic Energy Sciences (BES) Offices at DOE sponsor research and development (R&D) projects in this area. To facilitate progress, the project participants, including the institutions listed in Table 1, have formed a national Working Group on PEC hydrogen production, bringing together experts in analysis, theory, synthesis and characterization from the academic, industry and national laboratory research sectors.

Table 1. Some Participating Institutions in the DOE PEC Working Group.

University of Hawaii at Manoa
University of Nevada Las Vegas (UNLV)
University of Toledo
University of California, Santa Barbara (UCSB)
University of Nevada Reno
Caltech University
Stanford University
Colorado State University
National Renewable Energy Laboratory (NREL)
Intematix Corporation
MVSystems Incorporated
Midwest Optoelectronics
University of Louisville
Lawrence Livermore National Laboratory (LLNL)

This Working Group faces a daunting challenge. The complex set of requirements for an effective PEC material framing this challenge include: (1) efficient absorption of photons in the

178

solar spectrum; (2) sufficient photo-induced potentials to thermodynamically split water; (3) high collection efficiency of photo-induced carriers; (4) long-term stability in aqueous electrolytes; and (5) favorable kinetics for the electrode gas-evolution reactions. Moreover, a practical PEC material system must be low cost, using inexpensive materials fabricated by techniques compatible with large-scale manufacturing. In response to the challenge, the PEC Working Group is building a specialized *tool-chest* including state-of-the art techniques in materials theory, synthesis, characterization and analysis to facilitate research progress, and to help inspire fundamental breakthroughs. On a growing scale, the Working Group is successfully applying its tool-chest to investigate a broad spectrum of promising PEC materials classes. This paper gives a brief overview of the current DOE research approach to PEC hydrogen production research, and key Working Group activities.

MOTIVATION AND APPROACH

PEC hydrogen production based on the direct use of solar energy to split water is attractive among the solar-to-hydrogen (STH) conversion technologies because the efficient conversion can be achieved at low operating temperatures using cost effective thin-film and/or nano-particulate materials. PEC R&D efforts to develop enhanced semiconductor materials, devices and systems are moving forward, additionally benefiting from strong synergies with contemporary research efforts in the Photovoltaics (PV) and nano-technology communities.

Recognizing the long-term potential for practical STH production, DOE supports development of advanced PEC material systems. The overarching research approach integrates available state-of-the-art theoretical, synthesis, and analytical techniques to identify and develop the most promising material classes to meet the challenges of efficiency, stability, and cost. The motivation for this approach has been clearly outlined in the DOE Multi-Year Program Plan (MYPP) [4], detailing the fundamental technical barriers and philosophies for overcoming these barriers. The specific short-term and long-term research goals for PEC efficiency, taken directly from the DOE MYPP, are shown in Figure 1.

Table 3.1.10. Technical Targets: Photoelectrochemical Hydrogen Production[a]					
Characteristics	Units	2003 Status	2006 Status	2013 Target	2018 Target[b]
Usable semiconductor bandgap[c]	eV	2.8	2.8	2.3	2.0
Chemical conversion process efficiency (EC)[d]	%	4	4	10	12
Plant solar-to-hydrogen efficiency (STH)[e]	%	not available	not available	8	10
Plant durability[f]	hr	not available	not available	1000	5000

179

aThe targets in this table are for research tracking. The final targets for this technology are costs that are market competitive.
bTechnology readiness targets (beyond 2018) are 16% plant solar-to-hydrogen (STH) efficiency and 15,000 hours plant durability.
cThe bandgap of the interface semiconductor establishes the photon absorption limits. Useable bandgaps correspond to systems with adequate stability, photon absorption and charge collection characteristics for meeting efficiency, durability and cost targets.
dEC reflects the process efficiency with which a semiconductor system can convert the energy of absorbed photons to chemical energy [based on air mass 1.5 insolation] and is a function of the bandgap, IPEC and electronic transport properties. A multiple junction device may be used to reach these targets.
eSolar-to-hydrogen (STH) is the projected plant-gate solar-to-hydrogen conversion efficiency based on AM (Air Mass) 1.5 insolation. Both EC and STH represent peak efficiencies, with the assumption that the material systems are adequately stable.
fDurability reflects projected duration of continuous photoproduction, not necessarily at peak efficiencies.

Figure 1: Technical targets for PEC hydrogen production from DOE's MYPP

It is important to stress that PEC Hydrogen Production has already been successfully demonstrated on the laboratory scale. High STH efficiencies, between 12-16%, have been demonstrated for limited durations in devices based on expensive high-quality crystalline semiconductors. The most notable example is the III-V tandem $GaAs/GaInP_2$ cell [5, 6]. In addition, lower STH efficiencies, in the 3-5% range have been demonstrated in devices based on lower priced thin-film semiconductor materials. Multi-junction devices, for example using WO_3 films as a PEC top-junction, have been reported in this performance category [7, 8]. To achieve practical PEC Hydrogen Production, new semiconductor materials systems with both high performance and low cost are needed.

One specific approach is the further development of the traditional PEC semiconductor thin-films and nano-structures for higher efficiencies. Examples include improvements to iron-oxide and tungsten trioxide. Another approach is the adaptation of efficient PV semiconductor thin-films and nano-structures for effective use in PEC. This includes, for example, copper chalcopyrites and amorphous silicon compounds. Other innovative approaches include the development of entirely new materials classes, such as quantum-confined WS_2 and MoS_2 nanoparticle systems; and the development of breakthrough synthesis technologies to reduce the cost of high-performance crystalline semiconductors, such as $GaAs/GaInP_2$. Future progress in all these approaches will be integrally tied to the DOE PEC Working Group's continued development and deployment of its tool-chest, and continual feedback among the theory, synthesis and characterization efforts.

PEC WORKING GROUP ACTIVITIES AND PROGRESS

The success of PEC hydrogen production relies on meeting the significant technical challenges in developing new materials systems. Materials properties, including bulk and surface characteristics, are the key to efficient STH conversion and coordinated, collaborative materials R&D activities are essential. To foster collaboration, the DOE PEC Working Group has initiated

Task Forces to coordinate important PEC research activities. While some of the collaborative task forces center on the R&D of specific PEC materials classes, others focus on critical activities to advance the supporting science and technologies in the PEC tool-chest. Important activities in the latter category include:

–   Development of standardized testing and reporting protocols for evaluating candidate PEC materials systems on a level playing-field. In the past, the lack of standardized conditions and procedures for reporting PEC results has greatly hampered research progress across the board. To date, the *Standardized Testing Task Force*, which is being coordinated by NREL, has made significant initial progress, recently drafting eighteen detailed testing protocol documents that are under continued refinement for near term publication [9].

–   Development and refinement of techno-economic analyses of PEC hydrogen production systems incorporating performance and processing cost feedback from the broader materials R&D efforts. The objective is to provide a basis for evaluating the long-term feasibility of large-scale PEC production technologies in comparison with other renewable approaches. Techno-economic analysis activities with the Working Group have been coordinated through Directed Technologies Incorporated *(DTI)* [10].

–   Development of advanced characterization techniques to enhance understanding of PEC materials and interfaces and promote breakthrough discoveries [11]. The materials characterization efforts employ the most advanced microstructural, optoelectronic, and electrochemical characterization techniques available to paint a comprehensive picture of the materials properties in relation to PEC performance. Example techniques include X-ray photoelectron spectroscopy (XPS), ultraviolet photoelectron spectroscopy (UPS), Auger, Inverse photoemission spectroscopy (IPES), in *ex-situ* as well as new, advanced *in-situ* methods. Figure 2 shows the materials characterization facility developed at *UNLV*, a cornerstone of the PEC research characterization activities.

Figure 2: Advanced solid-state / surface characterization tool at UNLV

– Development of new theoretical models of PEC materials and interfaces critical to the design and engineering of new semiconductor systems [12, 13]. For PEC semiconductor materials, the effects of impurity incorporation and other asymmetries on the band structures need to be calculated using both traditional and enhanced density-functional theory algorithm. Developing these sophisticated models of band states and bandgap, including effects of surface, interfaces and grain-boundaries, has been initiated within the PEC Working Group at NREL and LLNL.

– Development and implementation of innovative synthesis techniques to facilitate the PEC materials discovery process [14, 15] represent important Working Group research activities. A broad spectrum of chemical, electrochemical and physical deposition methods are being employed to tailor material compositions and properties at UH, UCSB, and MVSystems, Incorporated, for example. Additionally, rapid throughput combinatorial methods based on the different synthesis routes are being explored. Innovative synthesis routes can make or break the viability of a semiconductor system, a fact well-appreciated by the PV community.

– Development of standardized screening procedures for the "Up-Selection" and "Down-Selection" of DOE supported PEC materials classes. The screening procedures are being developed in close coordination with the standardized testing, advanced characterization, and materials-theory activities. A hierarchy of screening protocols, ranging from top level device performance to detailed measurement of performance limiting materials and interface properties, is being formulated within the Working Group, as illustrated in Figure 3. Standard and consistent application of the best

182

available theoretical and experimental techniques to direct research efforts is the objective of this important Working Group activity. Examples of current screening equipment at UH and at NREL are shown in Figures 4 and 5, respectively.

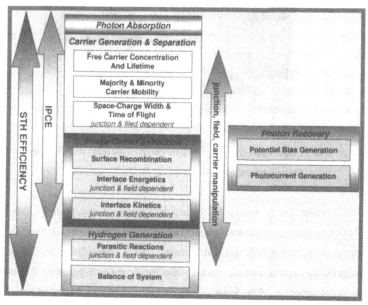

Figure 3: Block diagram of PEC Screening Hierarchy under development

Figure 4: Quantum Efficiency tool at UH useful in the standardized screening process for PEC materials

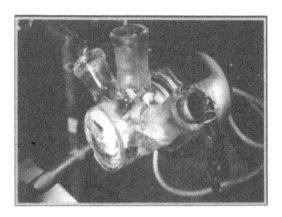

Figure 5: Solar-to-hydrogen efficiency measurement apparatus at NREL useful in the standardized screening process for PEC materials

In the PEC Working Group, continued feedback between the theory, synthesis, characterization and analysis is providing fundamental insights needed to promote technical breakthroughs in a broad spectrum of promising PEC materials classes. The Working Group has initiated a *White Paper* approach to the organization and tracking of scientific research progress. For each material class under investigation, a white paper is maintained as a living document which concisely summarizes the benefits, barriers, research status and approaches for addressing the barriers, all closely tied to the new advancements in theory, synthesis and characterization concurrently under development. The white papers also maintain a record of active research participants as well as a database of references documenting past achievements. To date, white papers have been drafted for a number of key focus materials classes including (but not limited to):

– Tungsten-oxide and related modified compounds - Tungsten oxide, particularly in thin-film and nano-particle forms, has been a workhorse in photoelectrochemical applications for years. It is inexpensive and stable, but its high bandgap (~2.6 eV) is limiting to PEC performance. Photocurrent densities of approximately 3 mA/cm^2 have been achieved [8, 16, 17], with STH efficiencies over 3% in tandem configurations. To break the performance barrier, current research is focused on reducing bandgap through ion incorporation into the WO$_3$ structure [13, 18], and further integration in multi-junction devices.

– Iron-oxide and related modified compounds - Iron-oxide is abundant, stable, inexpensive and has a near-ideal bandgap (~2 .1 eV) for PEC applications. Unfortunately, it's poor absorption, photo-carrier lifetime and transport properties have been prohibitive to practical water-splitting. Current research to overcome these barriers have been encouraging, with recent progress in thin films [19, 20, 21, 22] and nano-structured materials [23]. Iron oxide in tandem configurations may also be of interest for practical solar water-splitting.

− Amorphous silicon compounds, including silicon carbides and nitrides - Amorphous silicon compounds have recently demonstrated interesting performances in PEC applications [24, 25, 26, 27, 28, 29]. The progress of this material class in PEC applications has benefitted from decades of research in the PV community. Technical barriers remain in PEC stability and interface proportion, and electrolyte and surface modification studies could help overcome these barriers. With material and interface improvements, monolithically fabricated multi-junction devices using amorphous silicon compound films have practical appeal for PEC water splitting.

− Copper chalcopyrite compounds - Copper chalcopyrite thin films are among the best absorbers of solar energy. As a result, chalcopyrite alloys formed with copper and gallium, indium, sulfur and selenium have been widely characterized in the PV world [30, 31]. A great advantage of this material class for PEC applications is the bandgap tailoring based on composition, with bandgaps ranging from 1.0 eV in $CuInSe_2$ to 1.6 eV in $CuGaSe_2$, and up to 2.43 eV in $CuGaS_2$ [32, 33]. The $CuGaSe_2$ bandgap is attractive for PEC applications. Photo-current densities exceeding 13 mA/cm^2 have been demonstrated with this material [34] in biased PEC cells. Stability, surface kinetics and surface energetics remain as barriers, but if research can successfully address these, high STH efficiency could be achievable in low-cost thin-film copper chalcopyrite systems.

− Tungsten- and molybdenum- sulfide nano-structures - As bulk materials, tungsten- and molybdenum-sulfides are excellent hydrogen catalysts, but their bandgaps (below 1.2 eV) are too low for PEC water-splitting. Quantum confinement using nano-structuring, however, can increase the bandgap up to 2.5 eV. Current studies in nanostructured MoS_2 are focused on stable synthesis routes and integration of the nano-structures into practical bulk PEC devices [35].

− III-V semiconductor classes - High-quality crystalline semiconductor compounds of gallium, indium, phosphorous and arsenic have been studies for decades [36, 37]. In PEC experiments to date, STH efficiencies between 12 and 16 percent have been demonstrated in $GaInP_2$ /GaAs hybrid tandem photocathodes [5, 6]. High cost and limited durability are the barriers to practical PEC hydrogen production, and breakthroughs in synthesis and in surface stabilization are being pursued.

Although there is still much work ahead for achieving high-performance low-cost PEC hydrogen production, research in these promising candidate materials, among others, has seen significant recent progress. The reader is referred to the cited references for detailed progress in the

185

individual materials categories. The PEC Working Group's efforts to develop and deploy its theory/synthesis/characterization/analysis tool-chest has contributed significantly toward this progress. Continued Working Group growth, and extended international collaborations are being cultivated by the International Energy Agency's Hydrogen Implementation Agreement's Annex-26 on PEC Hydrogen Production promises more significant progress in the near future.

CONCLUSIONS

The US Department of Energy's Working Group on PEC hydrogen production has taken a collaborative approach in the R&D of novel PEC material systems. This approach, incorporating a broad spectrum of state-of-the-art techniques in theory, synthesis, characterization and analysis, is proving invaluable in the identification and development of the most promising materials for practical PEC hydrogen production. Continued Working Group efforts in conjunction with expanded international collaborations are expected to greatly facilitate the discovery and optimization of material systems and devices capable of meeting the DOE PEC hydrogen production targets.

ACKNOWLEDGEMENTS

The authors would like to acknowledge the contributions of the US Department of Energy, and the members of its Working Group on Photoelectrochemical Hydrogen Production. Special thanks are extended to the Principal Investigators within the Working Group, including Dr. Clemens Heske (UNLV), Drs. John Turner and Mowafak Al-Jassim (NREL), Dr. Eric McFarland (UCSB), Dr. Arun Madan (MVSystems Inc.), Dr. Tom Jaramillo (Stanford), and all others.

REFERENCES

1. Rifkin J. *The Hydrogen Economy : The Creation of The Worldwide Energy Web and The Redistribution of Power On Earth*. JP Tarcher/Putnam: New York, 2002.
2. Fujishima A, Honda K. Electrochemical Photolysis of Water at a Semiconductor Electrode. *Nature* 1972; **238**: 37–38.
3. Rocheleau R, Miller E. Photoelectrochemical Production of Hydrogen: Engineering Loss Analysis. *International Journal Hydrogen Energy* 1997; **22**: 771-782.
4. US Department of Energy Efficiency and Renewable Energy. Hydrogen, Fuel Cells and Infrastructure Technologies Program – Multi-Year Research, Development and Demonstration Plan; 2007. http://www1.eere.energy.gov/hydrogenandfuelcells/mypp/
5. O. Khaselev, J.A. Turner, Science **280**: 425-427 (1998)
6. O. Khaselev, A. Bansal, J.A. Turner, Int. J. Hydrogen Energy **26** 127-132 (2001).
7. Graetzel M. Photoelectrochemical cells. *Nature* 2001; **414:** 338-344.

8. Miller EL, Marsen B, Cole B, Lum M. Low-Temperature Reactively Sputtered Tungsten Oxide Films for Solar-Powered Water Splitting Applications. *Electrochemical and Solid-State Letters* 2006; 9(7): G248-G250.

9. "Standardized Testing Procedures for the Evaluation of Photoelectrochemical Materials and Systems", Document under preparation for publication, (2009).

10. "Techno-Economics Analysis of Photoelectrochemical systems for Solar Hydrogen Production". DTI presentation, to be presented at the 2009 DOE Hydrogen Program AMR, Arlington VA (2009).

11. L. Weinhardt, M. Blum, M. Bär, C. Heske, B. Cole, B. Marsen, and E. L. Miller, "Electronic Surface Level Positions of WO3 Thin Films for Photoelectrochemical Hydrogen Production", I Phys. Chem. C, 112, 3078-3082 (2008).

12. M. N. Huda, Y. Yanfa, C.-Y. Moon, S.-H. Wei, and M. M. Al-Jassim "Density-functional theory study of the effects of atomic impurity on the band edges of monoclinic WO3" Physical Review B 77, 195102 (2008).

13. Y. Yan and S.-H. Wei, "Doping asymmetry in wide-bandgap semiconductors: Origins and solutions", Phys. Stat. Sol. B 245, 641 (2008).

14. T. F. Jaramillo, S-H. Baeck, A. Kleiman-Shwarscstein, K-S Choi, G. D. Stucky and E. W. McFarland, J. Comb. Chem, 7, 264-271 (2005).

15. M. Woodhous, G. Herman and B. A. Parkinson, Chem. Mater. 17, 4318 (2005).

16. B. Marsen, B. Cole, E. L. Miller, "Progress in sputtered tungsten trioxide for photoelectrode applications", International Journal of Hydrogen Energy 32, 3110-3115 (2007).

17. B. D. Alexander, P. J. Kulesza, I. Rutkowska, R. Solarskac and J. Augustynski, "Metal oxide photoanodes for solar hydrogenproduction", J. Mater. Chem. 18, 2298–2303 (2008).

18. B. Cole, B. Marsen, E. L. Miller, Y. Yan, B. To, K. Jones, and M. M. Al-Jassim, "Evaluation of Nitrogen Doping of Tungsten Oxide for Photoelectrochemical Water Splitting", J. Phys. Chem. C 112, 5213-5220 (2008).

19. E. L. Miller, D. Paluselli, B. Marsen, R. E. Rocheleau, "Low-temperature reactively sputtered iron oxide for thin film devices", Thin Solid Films 466, 307-313 (2004).

20. A. Duret and M. Graetzel, "Visible Light-Induced Water Oxidation on Mesoscopic a-Fe2O3 Films Made by Ultrasonic Spray Pyrolysis", Journal of Physical Chemistry B, 109(36), 17184-17191 (2005).

21. Y.-S. Hu, A. Kleiman-Shwarscstein, A. J. Forman., D. Hazen, J.-N. Park, and E. W. McFarland, "Pt-Doped α-Fe2O3 Thin Films Active for Photoelectrochemical Water Splitting", Chem. Mater. 20 (12), 3803–3805 (2008).

22. A. Kleiman-Shwarscstein, Y.-S. Hu, A. J. Forman, G. D. Stucky, and E. W. McFarland, "Electrodeposition of α-Fe2O3 Doped with Mo or Cr as Photoanodes for Photocatalytic Water Splitting", J. Phys. Chem. C 112 (40), 15900–15907 (2008).

23. A. Kay, I. Cesar and M. Graetzel, "New Benchmark for Water Photooxidation by Nanostructured α-Fe2O3 Films", J. Am. Chem. Soc. 128, 15714-15721 (2006).

24. J. Hu, F. Zhu, I. Matulionis, A. Kunrath, T. Deutch, L. Kuritzky, E. Miller, and A. Madan, *"Solar-to-Hydrogen Photovoltaic/Photoelectrochemical Devices Using Amorphous Silicon Carbide as the Photoelectrode"*, 23rd European Photovoltaic Solar Energy Conference, Valencia, Spain, 1-5 September, 2008.

25. I. Matulionis, F. Zhu, J. Hu, J. Gallon, A. Kunrath, E. Miller, B. Marsen, and A. Madan, "Development of a Corrosion-Resistant Amorphous Silicon Carbide Photoelectrode for Solar-to-Hydrogen Photovoltaic/Photoelectrochemical Devices", SPIE Solar Energy and Hydrogen 2008, San Diego, USA, 10–14 August 2008.

26. F. Zhu, J. Hu, A. Kunrath, I. Matulionis, B. Marsen, B. Cole, E. Miller, and A. Madan, *"a-SiC:H Films used as Photoelectrodes in a Hybrid, Thin-film Silicon Photoelectrochemical (PEC) Cell for Progress Toward 10% Solar-to Hydrogen Efficiency"*, SPIE Solar Hydrogen and Nanotechnology 2007, San Diego, USA, 26–30 August 2007.

27. A. Stavrides, A. Kunrath, J. Hu, R. Treglio, A. Feldman, B. Marsen, B. Cole, E. Miller, and A. Madan, *"Use of amorphous silicon tandem junction solar cells for hydrogen production in a photoelectrochemical cell"*, SPIE Optics & Photons 2006, San Diego, USA, 13–17 August 2006.

28. S. Yae, T. Kobayashi, M. Abe, N. Nasu, N. Fukumuro, S. Ogawa, N. Yoshida, S. Nonomura, Y. Nakato, and H. Matsuda "Solar to chemical conversion using metal nanoparticle modified microcrystalline silicon thin film photoelectrode", Solar Energy Materials and Solar Cells 91, 224-229 (2007).

29. P.J. Sebastian, N.R. Mathews, X. Mathew, M. Pattabi, and J. Turner, "Photoelectrochemical characterization of SiC", International Journal of Hydrogen Energy 26 123-125 (2001).

30. I. Repins, M. A. Contreras, B. Egaas, C. DeHart, J. Scharf, C. L. Perkins, B. To, R. Noufi, "19.9%-efficient ZnO/CdS/CuInGaSe2 solar cell with 81.2% fill factor", Progress in Photovoltaics: Research and Applications 16, 235 (2008).

31. M. Bär, L. Weinhardt, S. Pookpanratana, C. Heske, S. Nishiwaki, W. Shafarman, O. Fuchs, M. Blum, W. Yang, and J.D. Denlinger, "Depth-dependent band gap energies in Cu(In,Ga)(S,Se)2 thin films", Appl. Phys. Lett. 93, 244103 (2008).

32. M. Bär, W. Bohne, J. Röhrich, E. Strub, S. Lindner,M.C. Lux-Steiner, and Ch.-H. Fischer, "Determination of the band gap depth profile of the penternary Cu(In(1-X)GaX)(SYSe(1-Y))2 chalcopyrite from its composition gradient", Appl. Phys. 96, 3857 (2004).

33. M. Bär, L. Weinhardt, C. Heske, S. Nishiwaki, and W. Shafarman, "Chemical structures of the Cu(In,Ga)Se2/Mo and Cu(In,Ga)(S,Se)2/Mo interfaces", Phys. Rev. B 78, 075404 (2008).

34. B. Marsen, B. Cole, E. L. Miller, "Photoelectrolysis of water using thin copper gallium diselenide electrodes", Solar Energy Materials & Solar Cells 92, 1054– 1058 (2008).

35. T. F. Jaramillo, K. P. Jørgensen, J. Bonde, J. H. Nielsen, S. Horch, and I. Chorkendorff, "Identifying the active site: Atomic-scale imaging and ambient reactivity of MoS2 nanocatalysts", Science 317, 100 – 102 (2007).

36. D. Crisp, A. Pathareb and R. C. Ewell (2004). "The performance of gallium arsenide/germanium solar cells at the Martian surface". *Progress in Photovoltaics Research and Applications* 54 (2): 83–101 (2004).

37 T. G. Deutsch, C. A. Koval, and J. A. Turner, "III-V Nitride Epilayers for Photoelectrochemical Water Splitting: GaPN and GaAsPN", J. Phys. Chem. B 110, 25297-25307 (2006).

Mater. Res. Soc. Symp. Proc. Vol. 1171 © 2009 Materials Research Society 1171-S09-10

Effect of Modified Titanium Dioxide on Rheological Behavior of Dental Porcelain Slurries for Rapid Prototyping Applications

Dongbin Zhu[1], Anping Xu[1], Yunxia Qu[1] and Jinsheng Liang[2]

[1]School of Mechanical Engineering, Hebei University of Technology, Tianjin, 300130, China
[2]Institute of Power Source & Ecomaterials Science, Hebei University of Technology, Tianjin, 300130, China

ABSTRACT

This study investigates bactericidal activities of dental porcelain contained silver/cerium/titania against *streptococcus mutans* in different irradiations, then the effect of modified titanium dioxide on rheological behavior of dental porcelain slurries is studied for rapid prototyping applications. The results show that the silver/cerium/titania has much higher antibacterial efficiency than that of pure TiO_2 either in the room light or under the dark. It is potential to apply this photocatalyst in the dental materials for preventing human teeth from caries because the oral cavity is mostly under dark, partly in the visible light. The modified titanium dioxide not only has a antibacterial effect, but also can improve the rheological behavior of dental porcelain slurries for rapid prototyping applications.

INTRODUCTION

Since Honda and Fujishima discovered the photocatalytic splitting of water on TiO_2 electrodes in 1972 [1-3], considerable effort has been devoted to developing highly active oxide semiconductor photocatalysts for their wide application in solar energy conversion and environmental protection. In recent years much attention has been paid to using TiO_2 as the antibacterial materials [4, 5]. However, the performance of TiO_2 requires further enhancement due to its narrow light response range as well as its low quantum efficiency [1-3, 6]. Consequently, numerous works focused on enhancing the photocatalytic performance of TiO_2 by doping, metal deposition, surface sensitization, coupling of composite semiconductors, and other methods [2, 7].

In previous research, a silver/cerium/titania composite material was prepared and the modified titanium dioxide has higher antibactericidal efficiency than pure titania, which arisen from not only the antibactericidal effect of the silver but also the synergistic effects of silver and cerium in the band gap narrow and the response to visible light, even under dark, and the suppressed the electron and hole recombination, and resulted in the enhanced antibactericidal performance [7].

It is potential to apply this photocatalyst in the dental materials for preventing human teeth from caries because the oral cavity is mostly under dark, partly in the visible light.

Solid freeform fabrication (SFF) is an automated manufacturing process that builds 3D complex-shaped structures layer-by-layer directly from CAD data without part-specific tooling and human intervention [8]. The development in SFF has offered opportunities to fabricate complex-geometry components and systems with highly integrated and self contained multi-functions that could not be easily fabricated using conventional approaches [9-12]. An example of this novel manufacturing process is the dental restoration using a selective slurry extrusion (SSE) based technique of SFF [13]. A green tooth can be produced by this method directly from a CAD digital model. The automation of the process offers opportunities to reduce the labor cost and increase the dental restoration rate.

One of the key issues in the SSE process is to deliver porcelain slurry to the desired location precisely with controlled shape. The rheological behavior of the porcelain slurry plays an important role in the shape of the extrudate. However, to the best of our knowledge, there are no literatures reported about the effect of modified titanium dioxide (silver/cerium/titania) on rheological behavior of dental porcelain slurries for rapid prototyping applications. In this research, the bactericidal activities of dental porcelain contained silver/cerium/titania are firstly investigated against *streptococcus mutans* in the dark, in the room light and under the UV light, then the effect of modified titanium dioxide on rheological behavior of dental porcelain slurries was studied for rapid prototyping applications

EXPERIMENT

The main chemical compositions of the dental porcelain powder are nano-ZrO_2 and modified-TiO_2. In which, the modified-TiO_2 are silver/cerium/titania composite materials prepared according to the reference [7]. Figure 1 shows the TEM micrograph of the dental porcelain powder obtained by a Philips Tecnai-20 TEM (Amsterdam, Holland).

The porcelain slurries were prepared by dispersing the dental porcelain powder described above in de-ionized water with a solid loading of 50 vol.%. An electromagnetic stirrer was used for the initial dispersion of the powder in the slurry. Then uniform and stable slurries were obtained by ball milling at a rate of 600 rpm for 24 h in a nylon resin coated jar with ZrO_2 balls as a milling media.

The *streptococcus mutans* was purchased from Tianjin medical university. The *streptococcus mutans* can be cultured on the dental porcelain slurries. A given amount of the bacterial solutions with known concentration was pipetted onto the slurry-coated glass plate and the uncoated glass plate as a control sample. Then, the bacteria were inoculated to sterile Luria Bertani (LB) medium and grown at 37 U over 12 h in the dark. The number of viable bacteria was determined by using the spread plate technique. The antibacterial efficiency of the films was measured through equation 1.

$$\eta = \frac{N_0 - N}{N_0} \times 100\%$$

(1)

Where, η means the antibacterial efficiency, N_0 the initial bacterial concentration and the viable bacterial concentration after treatment.

The zeta potentials of the porcelain slurries were measured by a laser light scattering electrophoretic technique (ZetaPlus, Brookhaven Instruments). The rheological properties of the slurries were analyzed through a MCR300 rotary rheometer. The pH of the slurries was determined using a PHS-3B precise pH meter (Shanghai precision & scientific instrument Co.,

LTD, China).

The porcelain slurries were extruded using a selective slurry extrusion (SSE) machine designed and constructed at Hebei University of Technology. The machine is shown in figure 2, which consists of the X-Y-Z positioning system and extrusion driver. The XYZ axes are driven by three stepping motors, respectively. The motion and position control was proved by a computer through a Baldor optimized multi-axis motion control card. A 736HPA high pressure dispense valve (EFD, U.S.A.) was mounted on the Z-axis and sample substrate was placed on the X-Y table.

Figure 1. TEM micrograph of dental porcelain powder.

Figure 2. The setup of selective slurry extrusion (SSE).

RESULTS AND DISCUSSION

The antibacterial efficiency is evaluated by the inhibition of bacterial growth of *streptococcus mutans*, which play one of most important roles in dental caries [14]. During metabolizing carbohydrates, *streptococcus mutans* produces organic acid, which can induce the demineralization of tooth surface and results in dental caries [15]. Figure 3 shows that antibacterial efficiency of the dental porcelain (contained 15 wt.% of silver/cerium/titania composite materials above mentioned) under the conditions of dark, room light and UV light.

On the glass, very weak antibacterial activities are obtained in all the conditions. It indicates that the UV/VIS light have little influence on the glass. The antibacterial efficiency of titania film is 2.6% in the dark, 20.6% in the room light and increases to 40.2% under UV light. Otherwise, the antibacterial efficiency of cerium/titania film is 3.2% in the dark, sharply increases to 60.2% in the room light and 75.5% under UV light. This indicates that the Ce in the cerium/titania film broadened the response light range and increased the photoactivity of titania film. As for the dental porcelain (contained 15 wt.% of silver/cerium/titania composite materials), an extremely high antimicrobial efficiency is also presented even in the dark. Moreover, its antibacterial efficiency in the room light and under UV light sharply increases to 66%, 98%, respectively, indicating that Ag nanoparticles in the silver/cerium/titania film are not only an antimicrobial but also an intensifier for photoactivity.

It is potential to apply this dental porcelain in the oral cavity because it can react to indoor light, even in the dark. However, the rheological behavior of dental porcelain slurries play an important role in the fabrication process of the artificial tooth via SSE. Therefore, in what fallows, we will study the effect of modified-TiO_2 on the zeta potential and shear rate of the dental porcelain slurries for rapid prototyping applications.

Figure 4 shows the zeta potential versus pH of the dental porcelain with and without modified-TiO_2. It was found that the isoelectrical point of the dental porcelain without modified-TiO_2 in deionized water is about pH 7.5. When the is below and above, the charge on the surface of porcelain powder is positive and negative, respectively. Ceramic powder dispersed in water can adsorb water molecules and form a hydration layer. Depending on the concentration of H^+ and OH^- ions in aqueous suspensions, the hydration reactions can take place.

At the isoelectrical point the number of the positive charge sites presented by zirconia powder H^+ ions is equal to the negative charge site of adsorption of OH^- ions, therefore, the net charge equals zero, and the particles in the suspension can be easily agglomerated. When more HCl and NaOH are introduced, the increasd ionic strength in the suspension compresses the thickness of the double electric layer, resulting in the increase of the absolute value of zeta potential of the powder.

It can be seen that after addition of modified-TiO_2 the isoelectrical point of dental porcelain shifted from pH 7.5 down to pH 6.5, however, the absolute value of the zeta potential is increased. In general, the dispersibility of ceramic powders in suspensions can be improved when the absolute value of the zeta potential is increased in good dispersion. The higher the absolute value of the zeta potential, the higher the charge density is on surface of powders and the larger the repulsion among particles.

Figure 5 presents the rheological property of the dental porcelain suspension with modified-TiO_2. With the increase in the shear rate there are the obvious decrease in the agglomeration role in the low shear rate area and the agglomeration was gradually broken when shear rate was further increased. Then the viscosity decreased slowly and eventually showed a constant value at very high shear rates. The rheological study also indicates that the modified-TiO_2 has an effect on the dispersibility of the dental porcelain powder, which is in accordance with the data of zeta potential.

As shown in figure 6, the green body of dental crown can be fabricated through SSE setup in figure 2. Moreover, this artificial tooth has an antibacterial effect for preventing human teeth from caries.

192

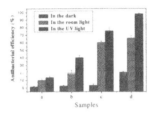

Figure 3. Antibacterial efficiency of the glass (a), titania (b), cerium/titania (c) and dental porcelain (d) under the conditions of dark, room light and UV light.

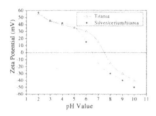

Figure 4. Zeta potential versus pH of the dental porcelain with and without modified-TiO$_2$

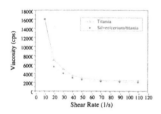

Figure 5. Rheological property of the dental porcelain with and without modified-TiO$_2$

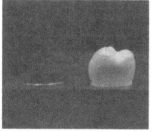

Figure 6. Green body of dental crown fabricating through SSE

CONCLUSIONS

In this paper, bactericidal activities of dental porcelain contained silver/cerium/titania against *streptococcus mutans* in different irradiations and effect of modified titanium dioxide on rheological behavior of dental porcelain slurries are studied. The modified titanium dioxide not only has an antibacterial effect, but also can improve the rheological behavior of dental porcelain slurries for rapid prototyping applications. The modified titanium dioxide can decrease the viscosity rate of dental porcelain slurry and the isoelectrical point of dental porcelain slurry from pH 7.5 down to pH 6.5.

ACKNOWLEDGMENTS

The authors would like to thank the "211" Project of Hebei University of Technology, and the Natural Science Foundation of Hebei Province under Grant Numbers: E2004000052 and E2006000039, for the financial support.

REFERENCES

1. A. Fujishima and K. Honda, *Nature* **238**, 37 (1972).

2. M.R. Hoffmann, S.T. Martin, W. Choi and D.W. Bahnemann, *Chem. Rev.* **95**, 69 (1995).

3. J. Yu, X. Zhao and Q. Zhao, *Thin Solid Films* **379**, 7 (2000).

4. D. Mitoraj,A. Janczyk,M. Strus, H. Kisch, G. Stochel, P.B. Heczko and W. Macyk, *Photochem. Photobiol. Sci.* **6**, 642 (2007).

5. J.Y. Choi, K.H. Kim, K.C. Choy, K.T. Oh and K.N. Kim, *J. Biomed. Mater. Res. Pt. B Appl. Biomate.* **80**, 353 (2007).

6. R. Asahi, T. Morikawa, T. Ohwaki, K. Aoki and Y. Taga, *Science* **293**, 269 (2001).

7. T. Zhang, D. Zhu, C. Wang and X. He, *Mater. Sci. Forum* **610-613**, 463 (2009).

8. J.J. Beaman, J.W. Barlow, D.L. Bourell, R.H. Crawford, H.L. Marcus and K.P. Mcalea, Solid Freeform Fabrication: A New Direction in Manufacturing, Kluwer Academic Publishers, MA, 1997.

9. L.C. Sun and L. Shaw, *Metall. Mater. Trans.* **30A** , 2549(1999).

10. K.J. Jakubenas, J.E. Crocker, S. Harrison, L.C. Sun, L. Shaw and H. Marcus, *Navel Res. Rev.* **L**, 51(1998).

11. J.W. Wang and L.L. Shaw. *J. Am. Ceram. Soc.* **89**, 346 (2006).

12. A.P. Xu and L.L. Shaw, *Comput. Aided Design*, **37**, 1308 (2005)

13. A.P. Xu, Y.X. Qu, J.W. Wang and L.L. Shaw, Design for Solid Freeform Fabrication of Dental Restoration. *Proceedings of the International Conference on Mechanical Engineering and Mechanics* in Nanjing, China, October 26-28, 2005, pp.1216.

14. W.N. Jessica and S. Gunnel, *Appl. Environ. Microbiol.* **73**, 5633 (2007).

15. T. Ooshima, Y. Osaka, H. Sasaki, K. Osawa, H. Yasuda, M. Matsumura, S. Sobue and M. Matsumoto, *Arch. Oral Biol.* **45**, 639 (2000).

Mater. Res. Soc. Symp. Proc. Vol. 1171 © 2009 Materials Research Society   1171-S10-03

Electronic Structure Properties of the Photo-Catalysts YVO4 and InVO4 Slab Systems With Water Molecules Adsorbed on the Surfaces

Mitsutake Oshikiri[1], Mauro Boero[2], Akiyuki Matsushita[1], and Jinhua Ye[1]

[1]National Institute for Materials Science, Tsukuba, Ibaraki, Japan

[2]Institut de Physique et Chimie des Materiaux, UMR 7504 CNRS-Universite Louis Pasteur 23 rue du Loess, BP 43, F-67034 Strasbourg Cedex 2, France

ABSTRACT

Electronic structure properties of a photo-catalyst slab system based on a material YVO_4 or $InVO_4$ which is sandwiched by water molecular layers have been investigated by first-principles calculation. As a result, we found a tendency that the band gap of the $InVO_4$ slab system sandwiched by water molecular layers was smaller than that of YVO_4 system while the band gap values of the bulk crystals of YVO_4 and $InVO_4$ are almost same. This result may provide us a good clue to understand the reason why the $InVO_4$ system can indicate a visible light response in photo-catalysis and the YVO_4 system can not.

INTRODUCTION

It is known that the YVO_4 photo-catalyst system [1] shows an excellent performance in production of both hydrogen and oxygen in the ultra-violet (UV) region, if with NiO_x co-catalyst, and $InVO_4$ system indicates visible light response in hydrogen generation up to the wavelength of approximately 600 nm [2, 3]. Yet, their catalytic properties, related to the electronic structure, are poorly understood. For example, the fact that the band gap values of the bulk crystals of YVO_4 and $InVO_4$ estimated by DFT-LDA calculations are almost identical (~ 3.3 eV), does not agree with the experimental results. In an attempt at unraveling this issue, we have investigated the effect of the water molecule existence and its adsorption to the surfaces of their photo-catalysts, on the electronic structure properties of the systems equilibrated around at room temperature by first-principles molecular dynamics simulations using a super cell model.

CRYSTAL STRUCTURE PROPERTIES

YVO4 crystal

In the YVO_4 crystal (a zircon-type crystal, the space group $I4_1/amd$) [4], each V is surrounded by four oxygen atoms forming a VO_4 tetrahedron (four-fold oxygen coordinated V; hereinafter called 4c-V) with an atomic distance of 1.71 Å between the V and the O, and each Y is surrounded by eight oxygen atoms forming YO_8 dodecahedron (8c-Y) with a Y-O distance of either 2.29 Å (for four of eight Y-O bonds) or 2.44 Å (for other Y-O bonds). The shortest V-V, Y-Y, O-O, and V-Y distances are about 3.9 Å, 3.9 Å, 2.6 Å and 3.1 Å, respectively. This YVO_4 crystal structure is characterized by the fact that every VO_4 tetrahedron is isolated by a YO_8 polyhedron. The conduction band minimum (CBM) of bulk YVO_4 is spanned by mainly V_3d atomic orbitals (~73 %), and the valence band maximum (VBM) is composed of O_2p (~83 %) [3].

InVO₄ crystal

On the other hand, the InVO₄ crystal (an orthorhombic system, the space group Cmcm) [5] includes 4c-V and 6c-In structures. The In-O distances of the InO₆ octahedron are almost identical at about 2.16 Å. Each VO₄ tetrahedron is isolated and not linked to other VO₄ tetrahedra. Therefore, the V-V distance is relatively large (4.05 Å). The characteristic V-O distance is 1.66-1.79 Å. The shortest O-O distance is ~ 2.7 Å and the In-In distance is 3.3 Å whereas the In-V distance is ~ 3.5 Å. The InVO₄ crystal has a laminated structure in contrast to the YVO₄ crystal. The CBM of bulk InVO₄ is spanned by mainly V_3d atomic orbitals (~59 %) and In_5s atomic orbitals (~19 %), and the VBM is composed of O_2p (~81 %) [3].

SIMULATION MODEL

YVO₄ catalyst system

In the present study, we focus on the (010) surface and we realize the surface using a slab included in a super cell. A slab surface including one set of 3c-V and 6c-Y and seven sets of 4c-V and 7c-Y structures was prepared to investigate the adsorption features depending on the oxygen coordination structures. All the atomic layers of the slab, except for the atomic layer of the other side surface, are allowed to be relaxed. The super cell contains four conventional unit cells of YVO₄ corresponding to a size of 2a × a × 2c, which includes 16 Y, 16 V, and 64 O atoms, and 35 water molecules. The super cell size is 2a × (a + water layers) × 2c = 14.23 Å × 11.94 Å × 12.57 Å (= 2.14 nm³) and the size of simulated surface, which is obtained by cleaving at the

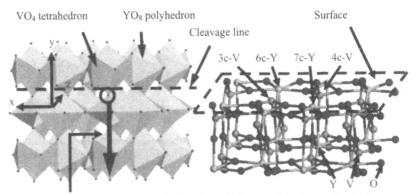

By shifting one of the corner O atoms of VO₄ on the surface by a translational vector (0, a, 0) of a conventional unit cell as indicated by the fat arrow, one set of 3c-V and 6c-Y structure can be exposed.

Crystal structure of YVO₄   Sketch of the slab surface

Figure 1. Crystal structure properties YVO₄ and the model surface employed in this study.

broken line in Figure. 1, is 2a × 2c = 178.94 Å². It is roughly identical to the size of one mono-molecular H_2O layer including 18.6 H_2O molecules in a normal room condition (common liquid water of 1.00 g/cm³ at 300 K, 1 atom). Therefore, the super cell realizes a model that approximately two mono-molecular H_2O layers are sandwiched by YVO₄ slab. The thickness (a + water layers) was optimized by a cell optimization process applying the Rahmam-Parrinello method with the cell parameters of 2a and 2c fixed. After the cell optimization process, the system was led to be at a thermal equilibrium state by applying constant temperature (300K) molecular dynamics by a velocity recalling algorithm and the band gap of the whole system was obtained in terms of Kohn-Sham energy calculation.

InVO₄ catalyst system

In the case of InVO₄ catalyst system, we prepared the (001) surface of InVO₄. The concerned top surface includes four 3c-V structures. Indium sites closest to the top surface are surrounded by six oxygen atoms (6c-In). The super cell includes two conventional unit cells of InVO₄ corresponding to a size of 2a × b × c as a slab part. The super cell size is 2a × b × (c + water layers) = 11.53 × 8.542 × 13.78 Å³ (= 1.36 nm³). A size of the simulated surface, which is obtained by cleaving at the broken line in Figure. 2, is 2a × b = 98.5 Å². It is roughly identical to the size of one mono-molecular H_2O layer including 10 H_2O molecules (10 × 3.1² = 96.1 Å²) in a normal room condition. The super cell has 8 In, 8 V, and 32 O atoms plus 30 H_2O molecules. The super cell realizes a model that approximately three mono-molecular H_2O layers are sandwiched by InVO₄ slab. After the cell optimization process, the system was led to be at a thermal equilibrium state at 300K and the band gap of the system was computed as in the former case.

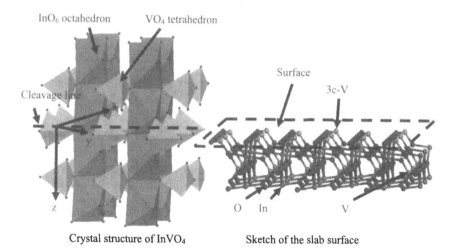

Crystal structure of InVO₄   Sketch of the slab surface

Figure 2. Crystal structure properties InVO₄ and the model surface employed in this study.

Computational frame work

First-principles dynamic simulations were performed within the Car-Parrinello scheme (CPMD) [6] using a Becke-Lee-Yang-Parr gradient corrected approach [7, 8]. The valence-core interaction was taken into account via norm-conserving Troullier-Martins pseudopotentials [9] for V, Y, In, and O atoms. For H, a Car-von Barth pseudopotential was used. In the case of V and Y, the use of semi-core states was needed in order to obtain a good description of both the geometry and the energetics. For In, non-linear core correction was included. The electrons of V 3s, 3p, 3d, 4s; Y 4s, 4p, 4d, 5s; In 5s, 5p; O 2s, 2p; H 1s were included in the valence electrons. Valence wavefunctions were expanded in plane waves with an energy cut-off of 80 Ry. A fictitious electronic mass of 1200 a.u. and an integration step of 5.0 a.u. ensured a good control of the conserved quantities.

RESULTS

Figure 2 (a) shows the snapshot of the catalyst surface of the YVO_4 slab with some adsorbed water molecules during molecular dynamics in a thermal equilibrium state at 300K. We can see the 3c-V structure disappears by the water molecule adsorption with forming H_2O-3c-V structure and some water molecules are adsorbed to the 7c-Y sites. On the other hand, Figure 2 (b) is the snapshot of the $InVO_4$ catalyst system in a thermal equilibrium state at 300K. Contrary to the former case, some of the water molecules are adsorbed dissociatively to the 3c-V sites and the 3c-V structures disappear by forming the H-O-3c-V structures. The band gap values of these systems are listed in the Table I. comparing to the band gap value of each bulk catalyst material. We can see the tendency that the band gap of the $InVO_4$ slab system sandwiched by water molecular layers is smaller than that of YVO_4 system while the band gap values of the bulk

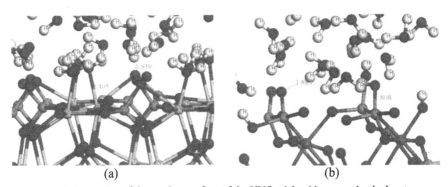

(a)   (b)

Figure 3. (a) A snapshot of the catalyst surface of the YVO_4 slab with some adsorbed water molecules during molecular dynamics in a thermal equilibrium state at 300K. (b) A snapshot of the catalyst surface of the $InVO_4$ slab with some adsorbed water molecules during molecular dynamics in a thermal equilibrium state at 300K. Numbers in the figures indicate atomic distances in angstrom unit.

Table I. Comparison of the band gap values (at Γ point).

	Bulk	Slab sandwiched by water molecular layers
YVO_4 system	3.28 eV	3.26 eV
$InVO_4$ system	3.31 eV	2.84 eV

crystals of YVO_4 and $InVO_4$ are almost same. This result indicates good correspondence to the experimental results with respect to the UV responsibility of the YVO_4 photo-catalyst system and the visible light responsibility of $InVO_4$ photo-catalyst system in the hydrogen generation. We also investigated the band gaps of some other kinds of slabs with different surface structure and varying the number of water molecules in the systems based on the YVO_4 and $InVO_4$ materials but the tendency with respect to the band gap was not changed. It may lead us to understand the reason why the $InVO_4$ system can indicate the visible light response in photo-catalysis and the YVO_4 system can not in the experiments so far.

CONCLUSIONS

Throughout our study, we have found a tendency that the $InVO_4$ system with water molecules adsorbed on a surface have a smaller band gap than that of YVO_4 with water molecules on a surface. The band gap of the $InVO_4$ slab with water molecules adsorbed on the surface is also smaller than that of bulk $InVO_4$. These theoretical research results suggest that we can make a photo-catalyst function in the longer wavelength region by a surface structure control in case using the $InVO_4$ material.

REFERENCES

1. J. Ye, Z. Zou, M. Oshikiri, T. Shishido, *Materials Science Forum*, **423-4**, 825 (2003).
2. J. Ye, Z. Zou, M. Oshikiri, A. Matsushita, M. Shimoda, M. Imai, and T. Shishido, *Chem. Phys. Lett.* **356**, 221. (2002)
3. M. Oshikiri, M. Boero, J. Ye, Z. Zou, G. Kido, *J. Chem. Phys.*, **117**, 7313 (2002).
4. J. A. Baglio, and G.. Gashurov, *Acta Cryst.*, **B24**, 292 (1968).
5. P. M. Touboul, and P. Toledano, *Acta Cryst.*, **B36**, 240 (1980).
6. R. Car, and M. Parrinello, *Phys. Rev. Lett.*, **55**, 2471 (1985); CPMD, Copyright IBM Corp. 1990-2001, Copyright MPI für FKF, Stuttgart, 1997-2004.
7. A. D. Becke, *Phys. Rev. A*, **38**, 3098 (1988).
8. C. Lee, W. Yang, and R. G. Parr, *Phys. Rev. B*, **37**, 785 (1988).
9. N. Troullier, and J. L. Martins, *Phys. Rev. B*, **43**, 1993 (1991).

AUTHOR INDEX

SUBJECT INDEX

Printed in the United States
By Bookmasters